1 MONTH OF
FREE
READING

at

www.ForgottenBooks.com

By purchasing this book you are
eligible for one month membership to
ForgottenBooks.com, giving you
unlimited access to our entire
collection of over 1,000,000 titles via
our web site and mobile apps.

To claim your free month visit:

www.forgottenbooks.com/free915938

ISBN 978-0-266-96032-4
PIBN 10915938

STATE OF NEW YORK

ANNUAL REPORT

OF THE

State Engineer and Surveyor

For the Year Ended June 30, 1919

ALBANY
J. B. LYON COMPANY, PRINTERS
1920

STATE OF NEW YORK

Office of the State Engineer and Surveyor

Albany, N. Y., *February* 16, 1920.

To the Legislature:

I beg to transmit herewith the State Engineer and Surveyor's annual report for the year 1919.

Respectfully,

FRANK M. WILLIAMS,

State Engineer and Surveyor.

[3]

TABLE OF CONTENTS

The report of the Gaging of Streams for 1919 is published as a supplemental
volume, or Vol. II.

REPORT

One of the duties of the State Engineer and Surveyor is to report to the Legislature each year on the activities of his Department. I submit herewith such report. For convenience the descriptive matter covers the calendar year of 1919. The financial tables and details of construction cover the fiscal year ended June 30, 1919. The State Engineer is a member of two constitutional boards, the Canal Board and the Land Board, and two permanent statutory boards, the Board of Canvassers and the Board of Equalization of Assessments. During the past calendar year the State Engineer served as a member of the following special boards or commissions: River Regulating Section of the Conservation Commission; Hospital Development Commission, Jamaica Bay-Peconic Bay Canal Board, Interstate Bridge Commission, the New York–New Jersey Bridge and Tunnel Commission, the Harlem River Improvement Board and the Gravesend Bay–Jamaica Bay Waterways Board.

For the past few years the construction of the Barge canal and its terminals has been the most important work carried on by the Department. With this work drawing to a close, the force has been materially reduced and the Department has been reorganized with the view of performing the engineering work necessary in the proper maintenance of the new canal system and to perform for other State departments work of an engineering character.

BARGE CANAL

During the navigation season of 1919 the canal has been in full operation and, while the traffic making use of the canal did not tax it to anywhere near its capacity, it demonstrated that from the purely construction standpoint the structures operated in accordance with the plans of the designers, and the failure to make a more extended use of the canal cannot be attributed to faulty design or operation. The uncompleted work, of which practically all is under contract, consists of the construction of additional bridges to span the channel, further protection to the banks and structures, widening the approach in the Rochester harbor and other miscellaneous work, all designed to improve navigating conditions. The new canal system was opened to navigation on May 15, 1918, and since that date has been in operation. The work herein referred to as uncompleted must not be classed as absolutely essential to the use of the canal, for such is not the case.

USE OF CANAL

In the season of 1918 boats were so scarce that the traffic was necessarily limited to the capacity of the available craft. In 1919 the number of boats navigating the canal was slightly increased. The Federal Government had built about seventy-five barges and several business corporations had their own boats. Some of the old canal boats were still in use, so that there were in operation about the equivalent of two hundred new barges.

There are some hopeful signs. Certain large companies have built and are using their own boats. Purchases of property adjacent to the canal have been made for the erection of plants or the storage of commodities. New articles have appeared on the list of canal freights.

It is not hard to discover ample reasons for this lack of sufficient traffic on the canal. The Government control, while it was expected to hasten by years the full development of canal traffic, has proved instead to be a hindrance. Although this control has not been nominally in force during 1919, nevertheless, through its regulation of traffic which might otherwise reach the canal and its ownership and operation of boats, its influence is still being felt. The lack of boats is easily explained. During the war the building of new craft by private companies was practically impossible. Although conditions have somewhat improved, still the high cost of labor and materials is holding back the building of canal barges. Moreover, the need of knowledge concerning both the canal and its advantages on the part of the shippers in general and also the lack of proper appreciation of its value account for much of the insufficient use. The process of building up traffic is necessarily slow. Traffic is peculiar in that once going in a given channel it often keeps to the same course irrespective of logic or reason.

BARGE CANAL TERMINAL, PIER 6, NEW YORK CITY

The freight-house has recently been completed and opened to traffic. The head-house contains the New York offices of the State Engineer and the Superintendent of Public Works.

CANAL MAINTENANCE

While the maintenance of the canal system does not come under the direct supervision of the State Engineer, it is necessary to furnish the engineering assistance required in the performance of the maintenance work. The canal has been completed and to be efficiently operated must be given proper maintenance. The equipment and force required to maintain the new canal differs greatly from that required on the old canal. The Superintendent of Public Works should be given the funds to properly equip his department to meet the problem now facing him in the maintenance of the new canal system.

FEDERAL CONNECTIONS — NOT YET MADE

The four main termini of the Barge canal system adjoin waters controlled by the Federal Government. Thus at the eastern and western extremities of the Erie branch are the Hudson and the Niagara rivers, respectively. At the northern end of the Oswego branch lies Lake Ontario. The Champlain branch terminates on the north in Lake Champlain, while its southern terminus is identical with the eastern terminus of the Erie branch.

The Government has improved the Niagara river to meet Barge canal requirements. At Oswego there was a stretch between the canal terminus and the lake harbor which the United States should have deepened, but which it entirely neglected. After waiting for some time for Congress to authorize this work, the State was obligated to open the channel, in order that the canal might not terminate in a dead end. Further harbor improvement at Oswego may with propriety be undertaken by the Federal authorities. During the past year Government dredging operations have been in progress in Lake Champlain — largely in straightening the tortuous channel at the southern end.

The Federal Government has provided an outlet in the Hudson of a size equal to the Barge canal, but for several years strenuous efforts have been made to induce Congress to deepen the channel north of the city of Hudson. This would allow canal barges to trans-ship their cargoes without making the 300-mile trip down the river and back. Also, it would help solve the difficult problem of a congested port at New York. The success attending somewhat similar undertakings in Europe argues in favor of this Hudson river deepening.

The State has spent a vast sum to improve its canals and it is only just that the United States should do its share and contribute at least a comparatively small amount to enable the State to reap its full measure of success. I can but renew my recommendation that your body take appropriate action to set the facts before Congress and ask its coöperation.

USE OF SURPLUS WATERS

The subject of conserving our natural resources, especially the water-power of our streams, has for several years been the theme of many a public utterance. Although we have talked about it for these many years and are nearly all agreed that something ought to be done, still we have actually accomplished but very little. In former reports I have given my views concerning the general subject of State water-powers and the particular feature of surplus canal waters, but I desire again to discuss these same topics.

With other State officials who have given the matter their attention, I believe that the State should adopt a strong, definite policy in treating the whole broad question of power development and flood control. Moreover, I am very strong in my opinion that prior to any final attempt to solve the problem this many-sided subject should be most thoroughly studied and also should be viewed from every angle, in order that the action which is eventually taken shall benefit with equal fairness both the State and the users of the power.

There is one phase of the subject, however, which demands special treatment, and which, moreover, should have immediate treatment. This is the utilization of power incidentally made available by Barge canal construction. Under the law the State at present is not permitted to dispose of surplus canal waters, and these waters are of such volume that from them there might be generated several thousand horse-power which is now going to waste. I believe that legislative action should be taken, and that without delay, to enable the State to lease this power. The canal interests should, of course, be safeguarded, but on the other hand

INTERIOR OF FREIGHT-HOUSE, PIER 6, NEW YORK CITY

Recently opened to traffic. Shed, 450 by 50 ft., head-house, 75 by 40 ft. Piers 5 and 6, operated as one unit, form the most important down-town canal terminal in New York.

the leases should be attractive enough to appeal to power-users. These power sites are situated on the line of the canal, in close proximity to the remarkable chain of thriving cities and villages that follows the State waterway. From the power thus supplied the State might receive a handsome revenue and at the same time industry would be benefited, since there would be conserved energy which otherwise would go to waste. Action on this matter need not interfere with any general water-power policy which the State may hereafter adopt in its treatment of the broad question of power development and flood control. I commend the subject to your careful consideration.

PROPOSED U. S. GOVERNMENT SHIP CANAL

The ship canal idea will not down.   Again this project has come to the fore. It has taken the form now of a proposal for the United States to give financial aid in the construction of a waterway of sufficient size to enable ocean-going vessels to reach the Great Lakes. Some of the Middle Western States are responsible for the passage in Congress of a measure which orders an investigation to determine what further improvement of the St. Lawrence river between Montreal and Lake Ontario is necessary for this purpose, together with an estimate of cost, and a report of recommendations concerning coöperation by the United States with Canada for this improvement. Immediately after the introduction of this measure and before final action could be taken, such bitter opposition arose that the advocates of an all-American route succeeded in having an amendment added, which authorizes a survey for a ship canal between the Great Lakes and the Hudson river.

Possibly the persistence of this ship canal idea is due to a human tendency, particularly an American tendency, to consider that whatever is biggest is necessarily best. It cannot be denied, too, that there is a fascinating glamour enveloping the thought of giant ocean ships penetrating to the heart of the continent and there exchanging the products which they have brought from the uttermost parts of the earth for the grain, the lumber and the ore of the vast Northwest. At the time of adopting the Barge canal as the State waterway policy there were certain fundamental prin-

ciples underlying the project, which determined the selection of
a barge rather than a ship canal and these principles were based
on careful observation and thorough study. I believe that the
reasoning was sound then and cannot see that any new factor has
entered the problem to warrant a different conclusion now.

Reduced to a single term, the reason for this selection was that
of cost — the greatest economy in cost of transporting cargoes.
Underneath this reason, however, there was a certain cause, and
this cause had to do with the types of vessels best fitted for particular kinds of navigation. Briefly, it was found that for the
highest enconomy in transportation special types of vessels are
needed, one for the ocean, one for the lakes and one for the canals,
and that no one type can supplant another in its proper waters
without suffering loss of economical efficiency. The further conclusion was reached, that it is not possible to combine these three
types into one vessel which will be as economical for the through
trip as to use the three existing types with two changes of cargo.
As I have said, these conclusions were based on actual observations, and it seems to me that unless they can be proved false,
sound business reasoning demands that which is based on practical economy and efficiency rather than something supported
chiefly by unproved theory or pleasing sentimentality.

If, however, a ship canal is to be built, the most logical route
is that known as the Oswego–Oneida Lake–Mohawk River route,
in connection with a canal between Lakes Erie and Ontario. This
route is practically that followed by the Oswego branch and the
eastern half of the Erie branch of the Barge canal. It is the route
favored in the report of the United States Deep Waterways Commission, which in 1897–1900 made exhaustive surveys and estimates for a ship canal between the Lakes and the Ocean. Incidentally it may be said that this report, because of the previous
report concerning the comparative costs of shipping by ship and
barge canals, fell substantially on deaf ears and has been of service
only in supplying data used in Barge canal construction.

I have said that the Oswego–Oneida Lake–Mohawk River route
is the logical course to be followed if a ship canal must be built.
Permit me to state the chief reasons for not coöperating in the
St. Lawrence project. It would take the control of the waterway

TRAFFIC ON THE BARGE CANAL

Several fleets just having been locked through the first lock at Waterford and starting on the west-bound trip through the canal. Most of the boats shown were built especially for Barge canal use and have dimensions considerably larger than those of the old canal-boats.

out of the hands of the United States, and this condition, I submit, is highly undesirable. Moreover, it would aid Canada far more than it would benefit any section of the United States, for it might divert from our own metropolis the commerce which has long been the bulwark of our growth and prosperity. Although our relations with Canada are most friendly, if the people of the United States realized that the scheme is chiefly in aid of a foreign power at the expense of our own success, I do not believe that they would for a moment look with favor upon the proposal.

There are certain physical conditions, also, which make the St. Lawrence route ill-advised. The river channel is hazardous, the coast along eastern Canada is foggy and extremely dangerous, and the port of Montreal is closed by ice during certain periods of the year. This latter condition is almost fatal to the hopes of any port that aspires to becoming a great ocean trans-shipping center. In contrast, New York with its port open the year around clearly shows its advantage.

Then, too, the return-load factor has always been against Montreal. Statistics show that the entire tonnage into Montreal is considerably less than the tonnage out of Montreal. The reason for this is to be found in the inability of Montreal and the territory tributary to it to absorb enough cargo to fill the ships needed to carry the products out. Here, again, New York stands in sharp contrast, for this port has usually been able to assure full cargoes both ways. This ability, of course, is due to the greater needs and purchasing power of the more highly developed hinterland in the United States. In fact, it would seem that shipping along the North Atlantic has grown up subject to the return-load factor and ports have attracted shipping largely in proportion to their ability to guarantee full loads both in and out. Of necessity this must be so. No elaborate terminals nor deep channels will ordinarily bring and hold shipping unless the territory dependent on the port can supply a cargo out and at the same time absorb a cargo in.

But aside from the relative merits of types of canal or routes of channel, it would seem that the only sensible thing to do now is to give the Barge canal a fair trial and it has not as yet had a fair trial. But something more is needed — something positive, aggressive and definite. And that something, it appears to me, is

a carefully-planned and vigorously-prosecuted campaign of educa-
tion, and this campaign should cover not only New York state,
but should be extended to all the territory adjacent to the Great
Lakes.   In view of the established fact that the cost of loading
and unloading is often equal to or even double the cost of actual
carriage, there is a wide field for improvement in terminal facilities.   If an adequate and proper program is carried out, shippers
will learn the advantages of shipping by water and will demand
and get suitable facilities.   The possibilities of water transportation are as yet little appreciated, but there is no reason why the
Barge canal should not fulfill its function and contribute largely
to the general prosperity of our State and Nation.

GRAIN ELEVATORS

One of the most promising fields for exploitation in the development of Barge canal traffic is in the grain trade.   When it is con-
sidered that from fifty to one hundred million bushels of wheat
are received in Buffalo each year and that at the close of navigation we find twenty times as much grain lying in Buffalo harbor,
awaiting shipment, as the canal has carried during the whole
season, we begin to realize the possibility of the grain trade.   And
when we perceive further that the vast grain belt lying around and
west of the Great Lakes covers an area of 1,250,000 square miles,
that it has a population of 30,000,000 and produces annually
5,000,000,000 bushels of grain, we may appreciate fully that, if
this class of canal traffic were increased as it might be, the maximum capacity of the canal would not be sufficient for its accom-
modation.   There is no valid reason why the canal should not
transport all of the grain it can handle.   This commodity is
admirably suited to water carriage; for the welfare of the world
the cost of shipping it should be as low as possible; the supply is
unlimited so far as the canal capacity is concerned, and other
carrying agencies are greatly in need of just such relief as the
canal could offer.   A further fact to remember in regard to the
grain belt is that the 30,000,000 people who produce these 5,000,-
000,000 bushels want to get their surplus products to the markets
of the world at the lowest possible transportation cost and that
they require for return cargoes the products of the world.   Thus

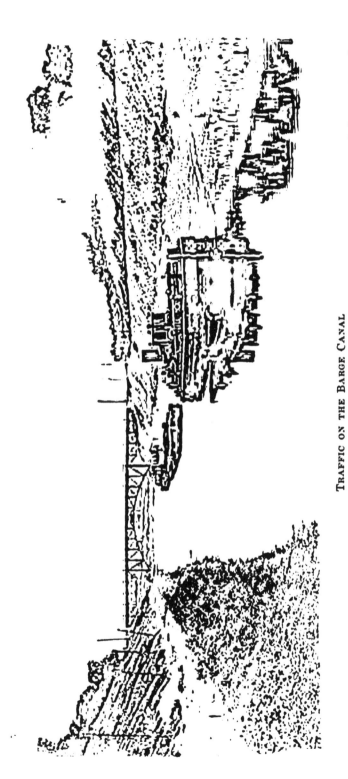

TRAFFIC ON THE BARGE CANAL

A fleet of four Federal-operated boats. Power boats of this type, capable of carrying partial cargoes, have recently been put on the canal. The view is at a point near Rochester where the deepest cut on the canal is located.

they will make our metropolis their port of export and import. It should be remembered also that in order to induce transportation companies to build and operate boats, we must make certain that there will always be full loads of grain available for east-bound trips. In this way, also, we may secure the vessels needful for other commodities and for our west-bound trade.

I have said that there is no valid reason why the Barge canal should not carry its due proportion of grain, but there is a reason why it cannot do this at present. It is plainly evident that the State must build and operate grain elevators.

The crying need is for an elevator at New York. At this port the two existing elevators are owned by railroad companies and there is no shadow of hope that these companies, competing with the canal for this traffic, will allow the canal to use their elevators. Even if they were favorably inclined, their facilities are insufficient for their own use. Such grain as now reaches New York by canal must of necessity, therefore, lie in the barges which bring it until ships are ready to receive the cargoes. Moreover, aside from the question of cost, the useless tying up of boats, which might otherwise be helping to build up canal traffic, is a condition which the State for its own good should hasten to remedy.

Canal officials have for some time perceived the need of a grain elevator at New York and plans have been made with a view of meeting this necessity. At Gowanus bay, Brooklyn, an area has been made available by depositing behind the new bulkhead wall the material from harbor dredging. Also I have planned and built the pier in such a way as to accommodate suitable grain-carrying apparatus. At this terminal there should be erected an elevator with working house and storage bins of ample capacity to meet the needs of the present and also of at least the near future, but so arranged that any required number of bins may easily be added as traffic increases.

At Buffalo there is not an actual lack of elevators, such as exists in New York, but for some reason the boatmen cannot seem to make satisfactory arrangements with the elevator owners to receive the amount of grain they desire and on terms which will make its transportation profitable. An elevator owned and operated by the State will doubtless remedy this trouble.

The necessity for a grain elevator at Oswego is almost as pressing as for one at New York. It is anticipated that the improvement of the Welland canal, which is now in progress, will to some extent cause lake boats to pass Buffalo and discharge their loads at ports at the eastern end of Lake Ontario — either Canadian or United States ports, as advantageous circumstances dictate. To participate in the grain trade following this course, the Barge canal must be equipped with a suitable elevator at Oswego.

This whole subject of State-owned and State-operated grain elevators is so important and, moreover, so essential to the material prosperity of our great canal system that it deserves your earnest and speedy attention. I therefore commend it to your consideration.

BARGE CANAL TERMINALS

The work of building and equipping Barge canal terminals has been steadily progressed during the year. This was the last portion to be undertaken in Barge canal construction, and although operations have been pushed as rapidly as circumstances have permitted, it will be the last part to be completed. Moreover, there are some phases of the terminal problem in which it is advisable to proceed cautiously. It may be predicted that a certain volume of traffic or a particular class of commodities will be handled at a given terminal, but experience may prove these opinions incorrect. Accordingly temporary freighthouses and partial installations of freight-handling machinery have been the policy in numerous cases. I have called these freighthouses temporary, but they have been well built and will last for years and the State will get full value for the money expended.

The work of the year has been the construction of certain docks, piers and warehouses and the installation of freight-handling machinery and electrical equipment. The New York city terminals have received the major part of our attention, but operations of more or less magnitude have been going on at other localities.

In the following paragraphs I shall take up the various terminals in order, beginning with those in New York city, and briefly describe what has been done.

ROOF CRANES AT PIER 6. NEW YORK CITY

These two 1½-ton roof cranes have been installed here. One is seen with its boom in a vertical position, ready for travel-ing along the roof, while the other is shown operating, a load being carried along its nearly horizontal boom.

At Pier 5 the repairs to the pier were completed. At Pier 6 the freighthouse was completed and the pier was paved. These two piers are operated as one unit and were turned over to the Superintendent of Public Works in a formal ceremony on October 14, 1919. The New York offices of the State Engineer and the Superintendent of Public Works are now located in the headhouse on Pier 6. There have been installed a complete lighting system for the offices and freightshed, and a power-distributing system for supplying current to semiportal revolving jib cranes, roof cranes, a portable conveyor, a tiering machine, capstans and a battery-charging equipment. Two 1½-ton roof cranes, five capstans, one tiering machine, one portable conveyor and a large number of hand trucks and trailers have been furnished and installed and two 3-ton semiportal revolving jib cranes are under contract and probably will be installed early in 1920. The power is brought underground to a switchboard in the freighthouse and thence distributed to the various circuits. From the same switchboard a circuit is carried to Pier 5 for operating the machinery there. Pier 5, for the present at least, is not to have any freightshed. It has been in constant use since the State took possession, even during the period when repairs were in progress.

At West 53d street the pier has been completed and the pier-shed is well under way. A contract has recently been awarded for building a headhouse. Work is also progressing in installing complete lighting, power-distributing and battery-charging systems. There have been delivered and installed capstans and a quantity of trailers and hand trucks. Two 3-ton semiportal revolving jib cranes, which are under contract, are expected to be installed early in 1920.

At Mott Haven the dockwall has been completed and the site has been paved. The interior of an existing brick building has been remodeled, so as to adapt it for use as a storage warehouse. Lighting and battery-charging equipments have been installed and hand trucks and trailers have been delivered.

At Flushing, work has been started in driving foundation piles for the dockwall and the frame freighthouse.

A contract for dredging and building a dockwall at Hallets Cove was awarded on November 26, 1919.

At Long Island City an existing bulkhead has been repaired, the site has been paved and a freighthouse is nearly completed. Complete lighting, power and battery-charging equipments have been installed.

At Greenpoint the new pier has been completed and alterations have been made in an existing concrete warehouse so as to adapt it for storage use. A shed on the new pier is nearly completed. A lighting system has been installed in the concrete warehouse, making it available for use the coming season. Provision is also being made for power and battery-charging installation for use in 1920. The battery-charging equipment and capstans have already been delivered and two semiportal revolving jib cranes are under contract. Also hand trucks and trailers have been delivered.

At Gowanus bay, work on the 1,200-foot pier has continued and a frame freighthouse has been built. The area for a depth of 60 feet behind the bulkhead wall has been paved, as has also the approach from Columbia street. The permanent freighthouse to be constructed on the pier will be 106 feet wide and 1,180 feet long. It differs from others on canal terminals in that it will have cargo masts along one side, these masts to be used in unloading ships by the burtoning method. Also the design is such as to permit the future addition of a conveyor gallery on the side of the house where the masts are placed. Should a grain elevator be erected, this addition of a conveyor gallery would be a part of that project. As I have said in former reports, from the beginning of terminal construction, those in authority have perceived the necessity of a commodious grain elevator in New York city, set apart especially for canal traffic. The site at Gowanus bay was best fitted for this elevator and all of the plans have been made with its possible erection in view.

Dredging operations to secure required depths of water have been in progress at three New York terminals — Piers 5 and 6, Long Island City and Greenpoint.

Apart from New York city the most important terminal construction has been at Buffalo. At Erie basin there is being erected a freighthouse, 80 by 500 feet in size, the freight section having a steel framework and the two-story office section, 80 by

BARGE CANAL TERMINAL AT LONG ISLAND CITY

At this place an existing bulkhead became serviceable by making certain repairs. The area has been paved and a freight-house built.

40 feet, having a reinforced concrete frame. The walls are of brick with artifical stone trim. This office section will house the Buffalo offices of the State Engineer and the Superintendent of Public Works. The building and the electric installation, which includes provision for lighting, power and battery-charging systems, is well advanced and is expected to be ready for use at the beginning of the 1920 navigation season. Tracks have been laid on both piers and railroad connections made. This terminal will be primarily the point of transfer between the lake craft and canal barges, but local traffic can also be handled.

At Ohio basin progress has been made in dredging the entrance channel and the Ohio street bascule bridge has been erected.

At Rochester a contract for a portion of the Court street viaduct, which forms an approach to the terminal, was awarded and nearly completed during the year. A contract for a temporary approach from Griffith street has recently been let. This contract includes also a 32 by 200-foot temporary freighthouse.

At Syracuse a frame freighthouse and timber derricks were completed. The house occupies the south pier. The work of constructing the harbor and piers was completed prior to a year ago.

At Utica the only work was in completing the pavement behind the freighthouse. The extension to the house had been finished before the first of the year.

At Amsterdam the second freighthouse and the pavement around the two houses were completed, although the greater part of this work was done in 1918.

At Albany the little work required to finish the freighthouse was completed.

At Whitehall the situation was similar to that at Albany and a small amount of work was needed to complete the freighthouse. The houses at Albany and Whitehall were alike in general character and were erected under the same contract.

At Oswego, track connection was made between the Delaware, Lackawanna and Western Railroad and the lake terminal.

The freight-handling machinery adopted at the various canal terminals include full-portal and semiportal traveling and revolving jib cranes of 3-ton capacity, 1½-ton traveling roof cranes, light and heavy duty stiff-leg derricks, package conveyors, burtoning devices, dock winches, capstans, carrying trucks and battery tractors. These are all electrically operated. The steam operated devices are locomotive cranes and small tractor cranes. There are also hand trucks and trailers, the latter being intended primarily for use in connection with battery tractors, but they may be operated by hand, if occasion demands.

WORK REMAINING IN COMPLETING TERMINALS

Another year has passed since I last called your attention to the fact that the funds available for constructing Barge canal terminals are all spent or obligated and that an additional sum is needed to bring these terminals to the degree of efficiency demanded by the needs of the canal, or even to the point contemplated when the original appropriation was made. The reason for this insufficiency of funds is apparent. Since the terminal work was begun the cost of labor and materials has advanced enormously. Under the act calling for compensation to contractors for increased costs due to war prices another extra toll was taken, and the early purchase of high-priced sites for New York city terminals has had its share in reducing the fund. Moreover, the original law did not provide for all of the terminals which properly should be built.

Several times I have set forth in my reports the imperative necessity of having canal terminals and especially of providing adequate freight-handling machinery. State-controlled docks and warehouses are needed to insure all shippers and boatmen an approach to the canal, but efficient handling devices are the chief means of reducing costs. To the uninformed, the ratio which the costs of loading and unloading bear to the cost of actual carriage is almost beyond belief.

The Barge canal must compete with the railroads and also with the Canadian canals. With all the obstacles it has to overcome — ignorance, indifference, prejudice, open opposition, sharp competition and the inertia of established commerce — it must be able to reduce shipping costs to a minimum. There is no field so fertile for reducing these costs as is that of terminal expenses. I recommend, therefore, that this subject receive your earnest attention.

MOVABLE DAM AT THE FOOT OF ROCHESTER HARBOR

Sector gate type of dam on the left, bridge type on the right. In removing the dam the sector gates are lowered into a recess and the bridge gates are raised and swung up under the bridge floor.

HUDSON RIVER TERMINALS

An appropriation for purchasing Barge canal terminal sites at Poughkepsie, Kingston, Newburgh and Yonkers was made by the Legislature of 1918. Preliminary surveys to secure data for acquiring the sites and for making plans were made during 1918. The lands have now been secured and studies have been made to determine the character of terminal construction needed.

The original Terminal Law made no provision for terminals on the Hudson between New York and Albany. By its action in 1918 the State has, in effect, pledged itself to the construction of these four terminals. The time is now ripe for the fulfillment of that pledge and I recommend that the appropriation necessary to proceed with this construction be made available.

INDUSTRIAL SITES ALONG CANAL

The recent purchase of three parcels of land near the Barge canal terminal at Syracuse by three large oil companies lends emphasis to a suggestion I have made in former reports concerning the availability of abandoned canal lands as sites for industrial plants. A few large companies, notably the General Electric Company and the Standard Oil Company, are appreciating the Barge canal and are making use of it. It would seem that only a knowledge of conditions is required to convince others of the value of water transportation. To such as are seeking industrial sites the abandoned canal lands present excellent locations. These are generally near the new canal, are not far from railroads and are in convenient proximity to power from surplus canal waters, the utilization of which seems destined to eventuate before long.

SCHENECTADY–SCOTIA BRIDGE

The bridge between Schenectady and Scotia, sometimes called the Great Western Gateway bridge, is in a sense a part of Barge canal construction, but it cannot really be regarded as such because of the design being more elaborate than canal interests demand. A large special appropriation, together with funds raised by Schenectady and Scotia, is needed in addition to the sum set aside from the Barge canal fund for a bridge at this location. Under authority of an act of 1917 I made careful plans and estimates of this bridge and reported my findings to the Legislature. By an act of 1919 construction along the line of these plans was ordered. Immediately after the passage of this act I reorganized a part of my force so as to get the work ready for letting with the least possible delay. It was decided to divide the work into four contracts. The first of these includes the abutments and approaches, exclusive of paving, at both Schenectady and Scotia; the second is for the piers below the undercoping; the third provides for the entire superstructure, while the fourth takes care of the paving.

The new bridge is to be situated about a half-mile west of the bridge which now forms the link between Schenectady and Scotia. It is to be of reinforced concrete arch construction, having 23 arches with spans ranging from 106 to 212 feet. The length of the concrete structure is 3,186 feet and the approaches bring the total length to 4,436 feet.

Bids were opened for the first contract last September, but unfortunately, because Scotia's share of the funds was not available, the possibility of making an award was delayed for about three months. At the end of that time the lowest bidder declined to accept the contract, claiming that prices had advanced and his options had been for thirty days and he could not perform the work for the amounts in his bid. The second lowest bidder was given an opportunity to take the work at his proposal, but he also declined, and now the contract is being readvertised for a second opening of bids. It seems probable that the new bid will exceed the original bid by many thousand dollars. The plans for the second contract are now completed.

BLUE LINE SURVEYS AND SALE OF ABANDONED CANAL LANDS

During the past year progress has been made in computing the blue line surveys and maps on which the abandonment and sale of old canal lands is based. The work of mapping these old canal lands has been principally confined to sections where there was a prospect of making a sale of the lands involved.

So far as this Department is concerned, this work has reached such a stage that the greater portion of the lands to be sold can be

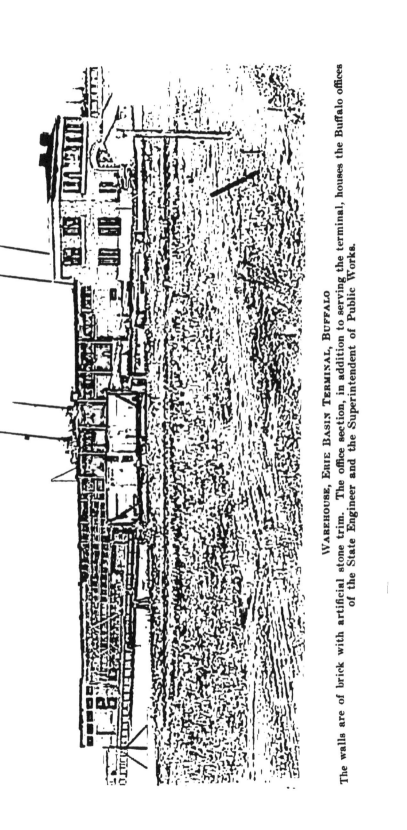

WAREHOUSE, ERIE BASIN TERMINAL, BUFFALO

The walls are of brick with artificial stone trim. The office section, in addition to serving the terminal, houses the Buffalo offices of the State Engineer and the Superintendent of Public Works.

disposed of during the coming year; this includes old canal lands in the cities of Utica, Syracuse and Rochester, from the sale of which the State should receive considerable revenue. From the sales made to date of January 1, 1920, embracing seven miles of old canal, the State has realized $426,818.56.

REORGANIZATION OF DEPARTMENT

At the time the Barge canal and terminal construction was under full headway over 90 per cent of the cost of maintaining this Department was paid out of bond moneys. As this work was brought toward completion the force was reduced. It is apparent that of the force remaining in the employ of the Department the greater portion of their time is devoted to maintenance of the canal and work other than construction. It is evident that this force should no longer be charged against a bond issue provided for construction only, so that the amount requested from the Legislature to run the Department has been based on the transfer of these employees to the payroll to which they should be equitably charged. The cost of maintaining the new canal will be much in excess of the cost of maintaining the old system, and with a complete understanding of the situation it is not surprising that so far as legislative appropriations are concerned, the cost of running the Department aside from construction is increased.

In the budget requests for the coming fiscal year, provision has been made to place the Department in a position to undertake such work as it may be called upon to perform as the engineering department of the State. The only employees charged to bond moneys will be those directly connected with the completion contracts on canals and terminals. If the new budget requests are granted, it will be unnecessary to make an appropriation to meet the engineering expense of each new activity undertaken by the State and this Department will be organized to furnish such engineering advice through the regular employees of the Department who are paid out of the regular departmental appropriation received from the Legislature.

During the past year this Department joined with the Highway Commission and the Conservation Commission in a request to the Civil Service Commission for a readjustment of the salary grades for engineers. With the approval of the Governor, the Civil Service Commission adopted these new grades and the same are now in effect. This provides for increases in the compensation of engineers, to which in my opinion they are justly entitled. I trust in considering the requests for appropriations for personal service the Legislature will look with favor to the payment to the engineers of the salaries as set forth in the new schedule which applies to these various positions.

APPENDED REPORTS

The usual reports and tables are hereto appended. First in order are the tables which give summaries of engineering expenses. These are followed by tables of contracts, both those completed during the fiscal year and those in force at its close. A table which summarizes, by years and canals, the work of constructing the Barge canal and its terminals completes the chief tabular appendices. Next in order are the reports of the three Division Engineers. They give in detail the accounts of nearly all the engineering and contract work that has been in progress. A detailed tabular statement of engineering expenditures accompanies each Division Engineer's report. Other reports covering activities of the Department are appended.

ACKNOWLEDGMENTS

During the past year the relationship existing between this Department and the Department of Public Works has been most cordial and I desire to acknowledge the assistance rendered and the spirit of coöperation shown at all times by the Superintendent of Public Works, Edward S. Walsh.

The employees of this Department have continued to render faithful and efficient service, for which I desire to express my thanks.

Respectfully submitted,
FRANK M. WILLIAMS,
State Engineer and Surveyor.

ENGINEERING EXPENSES FOR THE FISCAL YEAR
ENDED JUNE 30, 1919

TABLE OF CONTRACTS COMPLETED DURING THE
FISCAL YEAR ENDED JUNE 30, 1919

TABLE OF CONTRACTS PENDING JUNE 30, 1919

SUMMARY OF CONSTRUCTION WORK, BARGE CANAL
AND TERMINALS, BY YEARS

Hudson River Terminals

WORK	ACT		Division	Amount	Total
	Chap.	Year			
Hudson river terminals...........	555	1918	Eastern....	$4,816 58	$4,816 58

Bridge Designers, Engineers, etc.

WORK	ACT		Division	Amount	Total
	Chap.	Year			
Bridge designers, engineers, etc....	151	1918	Eastern....	$2,000 00	$2,000 00

Special Work

WORK	ACT		Division	Amount	Total
	Chap.	Year			
High street bridge, Cohoes.......	181 151	1917 1918	Eastern....	$820 98	
Schenectady–Scotia bridge.......	735 634	1917 1919	Eastern...	3,356 02	
Sea-wall, Orient–East Marion....	428	1918	Eastern....	739 52	
					$4,916 52
Glen creek improvement........ ..	341	1918	Middle....	$120 87	
Dive culvert, Rome	346	1918	Middle.....	100 89	
Lake street bridge, Geneva.......	351	1918	Middle.....	425 41	
Minetto bridge.................	716	1915	Middle.....	1,371 20	
Limestone creek improvement.....	751	1917	Middle.....	14 18	
Canandaigua lake dredging	756	1917	Middle.....	1,003 75	
Cowasselon creek dredging..	781	1917	Middle.....	419 15	
Whitesboro street bridge, Rome...	753	1917	Middle.....	1,324 60	
					4,780 05
Chadakoin river improvement....	758 728 181 644 624	1913 1915 1917 1919 1913	Western....	$555 05	
Elliott creek improvement.	728 181 760	1915 1917 1917	Western....	2,195 79	
Hertel avenue bridge, Buffalo.	761	1917	Western....	1,625 78	
Eighteen-Mile creek culvert, Lock-port....	181 626	1917 1917	Western....	55 80	
Griffin creek improvement, Cuba...	565	1918	Western....	397 40	
					4,829 82
Total......................	......	....	....	....	$14,526 39

Special Surveys

WORK	ACT		Division	Amount	Total
	Chap.	Year			
Blue line surveys..............	151	1918	Eastern....	$15,000 00	
Surveys for State Court of Claims .	151	1918	Eastern....	200 00	
Department surveys	151	1918	Eastern....	3,627 93	
State boundary...	151	1918	Eastern..	378 26	
Delaware–Schoharie county boundary line........................	559	1918	Eastern...	2,375 55	
Saratoga–Warren county boundary line.......	561	1918	Eastern...	2,476 80	
Ulster–Greene county boundary line {	562 600	1918 } 1919 }	Eastern....	3,682 69	
Land grants.....	151	1918	Eastern..	2,718 69	
Survey of lands under water.....	12	1918	Eastern...	912 64	
Jamaica Bay–Peconic Bay canal. . {	317 343	1917 } 1918 }	Eastern...	4,177 99	
Mill river survey................	427	1918	Eastern...	466 55	
Hydrographic survey............ {	181 151	1917 } 1918 }	Eastern....	2,982 93	
					$38,941 93
Blue line surveys, Erie canal... .	151	1918	Middle....	$17,116 14	
Blue line surveys, Oswego canal....	151	1918	Middle ..	266 13	
Survey for State Court of Claims, Erie canal	151	1918	Middle....	2,494 29	
Survey for Hospital Development Commission, Utica State hospital, Marcy division................	151	1918	Middle....	1,792 41	
					21,668 97
Blue line survey	151	1918	Western...	$10,500 00	
Survey for State Court of Claims .	151	1918	Western....	2,150 00	
Eighteen-Mile creek, Niagara county.....................	425	1918	Western....	2,333 74	
					14,983 74
Total......................	......	......	....	....	$75,594 64

Summary of Engineering Expenses for the Fiscal Year Ended June 30, 1919

DIVISION	Ordinary repairs to canals	Construction of Barge canal	Construction of Barge canal terminals	Hudson river terminals	Bridge designers, engineers, etc.	Special work	Special surveys	Totals
Eastern and head office..	$10,000 00	$299,219 46	$155,691 78	$4,816 58	$2,000 00	$4,916 52	$38,941 93	$515,586 26
Middle	9,984 92	75,790 17	18,765 19	....	....	4,780,05	21,668 97	130,989 30
Western.....	10,000 00	138,480 98	26,980 94	....	....	4,829 82	14,983 74	195,275 48
Totals....	$29,984 92	$513,490 60	$201,437 91	$4,816 58	$2,000 00	$14,526 39	$75,594 64	$841,851 04

TABLE OF CONTRACTS COMPLETED DURING THE FISCAL YEAR ENDED JUNE 30, 1919

Special Work

CONTRACTOR	Date of contract	Character of work	Division	Act Chap.	Act Year	Appropriation	Engineer's preliminary estimate	Contract price as modified by alterations	Final payment
Rosoff Engineering Co....	June 12, 1918	Constructing a concrete pavement on State reservation, Rockaway Point	Eastern	130	1917	*	$35,995 00	$48,126 25	$45,989 95
Rosoff Engineering Co....	Oct. 18, 1918	Completing sea-wall between East Marion and Orient	Eastern	428	1918	*	6,812 00	9,109 00	8,818 81
A. M. Hasell, Inc........	Jan. 3. 1918	Repairs to landing pier, Quarantine station, Hoffman island, New York harbor	Eastern			†	......	3,387 00	4,152 00
Anderson & Wheeler.....	June 26, 1918	Repairs to north pier, Quarantine station, Hoffman island, New York harbor	Eastern			†	......	......	4,562 02
M. Field..........	Sept. 9, 1918	Elimination of High st bridge, Cohoes	Eastern	{ 181 / 151	1917 / 1918 }	......	12,204 50	12,351 75	12,283 50
Walter S. Rae........	Oct. 25, 1917	Slinging a steel bridge over the Black River canal at Two street, Rome	Middle	753	1917	$15,000 00	12,085 00	11,872 30	11,142 86
Bert Provo..........	Nov. 30, 1917	Dredging and improvement of Cowasselon creek, between Canasta and Lakeport	Middle	781	1917	12,000 00	10,120 00	10,380 00	9,572 70
Larkin & Sangster........	Sept. 12, 1916	Constructing portions of a bridge over the Oswego river at ... (Part of Barge canal contract No. 99)	Middle	{ 716 / 181	1915 / 1917 }	50,000 00	44,088 15	42,988 15	§
W. F. Martens..........	May 6, 1918	Dredging harbor and repairing pier and break-water at Canandaigua	Middle	{ 756 / 760 / 181	1917 / 1917 / 1917 }	16,500 00	......	15,097 00	12,650 94
J. W. Hennessy, Inc......	April 18, 1918	Improvement of Ellicott creek, Erie county	Western			88,711 38	83,803 25	96,885 30	77,281 96
Lupfer & Remick.......	Mar. 15, 1918	Construction of a steel through-truss bridge over the old Erie canal at Hertel avenue, Buffalo	Western	761	1917	30,000 00	27,937 50	27,967 20	25,311 20
Geo. L. Maltby..........	Mar. 23, 1916	Improvement of Chadakoin river, Chautauqua county	Western	{ 758 / 728	1913 / 1915 }	100,000 00	89,252 25	92,074 25	¶

* Work done under an act which, in addition to making provision for the acquisition of land, appropriated $100,000.00 for highway improvement, for purposes of public defense.

† These contracts have been prepared and the work was supervised by this Department for the Health Officer of the port of New York. The funds were available from appropriations for the use of this officer.

§ Payment for work done was made under chapter 585, Laws of 1918. Part of contract No. 99 was in this payment and could not be separated.

¶ Contract finished under direction of Superintendent of Public Works.

Construction of the Barge Canal

Chapter 147, Laws of 1903; chapter 391, Laws of 1909; and amendatory laws

CONTRACTOR	Date of contract	Character of work	Division	Engineer's preliminary estimate	Contract price as modified by alterations	Final payment
Holler & Shepard	Aug. 31, 1914	Contract No. 1-A, Champlain canal — Hudson river. Northumberland to Fort Miller and Crockers Reef to Fort Edward	Eastern	$90,811 00	$141,340 20	*$211,604 61
H. S. Kerbaugh, Inc	Nov. 3, 1916	Contract No. 19-A, Erie canal — Redredging contract No. 19	Western	152,200 00	160,750 10	*236,943 42
Walsh Construction Co	Feb. 16, 1916	Contract No. 21-A, Erie canal — Completing Genesee river to contract No. 6	Western	115,700 00	428,475 54	*020,327 80
H. S. Kerbaugh, Inc	May 20, 1916	Contract No. 23-A, Erie canal — Completing King's Bend to Genesee river	Western	651,703 10	745,072 12	*079,919 99
Eastover Construction Co., Inc	Mar. 27, 1916	Contract No. 29-A, Erie canal — Completing Sterling creek to Herkimer-Oneida county line	Eastern	162,005 00	318,639 70	*371,408 20
Grant Smith & Co. & Locher	Feb. 24, 1913	Contract No. 42-A, Erie canal — Completing Herkimer-Oneida county line to Oriskany road	Middle	1,033,037 85	1,239,045 03	1,197,244 78
Scott Bros	Oct. 10, 1916	Contract No. 44-A, Erie canal — Completing prism near junction lock at New London	Middle	37,050 00	32,486 00	*56,242 37
Scott Bros	Feb. 25, 1916	Contract No. 46-B, Erie canal — Lock, dam, etc., at May's Point	Middle	314,040 72	283,676 97	269,398 41
State Highway Construction Co	Feb. 23, 1916	Contract No. 63-A, Erie canal — Completing canal, west line of Wayne county to King's Bend	Western	367,745 70	581,861 50	*387,777 24
Great Lakes Dredge and Dock Co	Jan. 15, 1916	Contract No. 73-A, Champlain canal — Completing canal from Northumberland to Stillwater	Eastern	432,045 00	505,160 67	*577,985 82
Mohawk Dredge and Dock Co	Oct. 22, 1917	Contract No. 83, Erie canal — Canal at Tonawanda and removing guard-lock and coffer-dam	Western	149,604 50	195,351 00	138,466 90
Jupfer & Remick	Mar. 9, 1917	Contract No. 84, Erie canal — Viaduct over Clyde river at Clyde	Western	83,984 50	83,876 00	*89,390 37
The Holington Co	Jan. 5, 1911	Contract No. 91, Erie canal — Hydro-electric power-plant at Crescent dam	Eastern	44,600 00	44,985 50	40,458 32
Tuft Construction Co., Inc	Nov. 24, 1916	Contract No. 98, Erie canal — Adams street lift-bridge, Lockport	Western	77,495 00	82,425 25	*83,380 98
Larkin & Sangster	Sept. 12, 1916	Contract No. 99, Oswego canal — Bridge over Oswego river at Minetto	Middle	117,170 75	115,080 75	*152,347 24
Chelsey, Earl & Heimbach, Inc	Mar. 8, 1917	Contract No. 122-A, Erie canal — Completing highway bridge near Little Falls	Eastern	52,717 00	67,377 00	81,716 95
M. Fitzgerald	Mar. 5, 1917	Contract No. 131-A, Champlain canal — Completing portion of a highway bridge at Schuylerville	Eastern	30,753 00	39,634 50	37,267 28

TABLE OF CONTRACTS COMPLETED DURING THE FISCAL YEAR ENDED JUNE 30, 1919 — (Continued)
Construction of the Barge Canal—(Continued)

CONTRACTOR	Date of contract	Character of work	Division	Engineer's preliminary estimate	Contract price as modified by alterations	Final payment
W. F. Das & Son	Mar. 8, 1917	Contract No. 141, Erie canal — Power-station at lock No. 29, Palmyra	Western	$41,166 50	$11,180 75	$27,457 82
W. F. Martens & Co. Inc	June 14, 1917	Contract No 144, Erie canal — Two concrete bridges over Red creek, Shocco Valley park, Rochester nr	Western	41,480 70	41,258 70	16,580 00
Chelsea, Earl & Heinbach, Inc	Aug. 28, 1917	Contract No 156, Erie canal — H way bridge nr Kylvan Beach	Middle	7,788 00	10,113 00	9,643 30
I. M. Idington's Sons, Inc	Mar. 27, 1917	Contract No. 159, Erie canal — Extending Ganargua c ed spillway and raising canal bank nearby	Western	30,464 00	43,258 5?	40,684 26
Lord Construction Co	Aug. 3, 1917	Contract No. 161, Erie canal — Electric motors and machinery for guard-lock s at Rochester	Western	5,972 00	15,867 35	15,750 20
Mk Dredge & Dock Co. Inc	Nov. 23, 1917	Contract No 165, Erie canal — Removing Montezuma aqueduct	Middle	81,530 00	160,943 00	145,798 26
Cleveland & Sons Company	Nov. 10, 1917	Contract No. 170, Erie canal — Junction lock at South Greece	Western	54,800 50	64,942 50	56,444 24
Lupfer & Remick	Mar. 15, 1918	Contract No. 172, Erie canal — Barrel bys and lamp-posts on the Seneca, Clyde, dve and Tonawanda rivers	Middle and Western	14,853 00	12,921 45	12,913 35
Dunbar & Sullivan Dredging Co	Mar. 15, 1918	Contract No. 180, Erie canal — Removing part of aqueduct at Rexford Flats	Eastern	17,840 00	15,958 00	16,153 54
Law Bars	Dec. 28, 1917	Contract No 181, Erie canal — Lining and water-proofing prism at Little Falls	Eastern	46,624 00	48,253 50	48,222 72
Donnell-Zane Co	Sept. 11, 1918	Contract No. 184, Erie canal — Aligning bridge, west Henrietta oad, Rochester	Western	6,850 00	5,915 25	5,504 53
Mohawk Dredge & Dock Co., Inc.	April 12, 1918	Contract No. 184, Erie canal — Excavating under N. Y. C. R. R bdge at Brew tan	Middle	7,200 00	9,480 00	9,562 95
Lupfer & Remick	Nov. 5, 1914	Contract M, (ica and Seneca cnl — Electrical and operating ent for lcks No. 1, 2, 3 and 4	Middle	1,067 00	191, 4600	190,274 64
The Sherman-Stalter Co	April 30, 1918	Contract R, tfa and Seneca canal — Completing unfinished work on Cayuga and Seneca canals	Middle	185,259 00	180,122 80	173,434 38
Smith Soper	Jan. 3, 1919	Contract U, Cayuga and Seneca canal — Repairing manholes of sewer in Benton creek, Seneca Falls	Middle	5,941 00	7,382 00	5,147 68

* Final payment made under chapter 585, Laws of 1918.   † Canceled May 7, 1919.

Special Work Connected with Barge Canal Construction

CONTRACTOR	Date of contract	Character of work	Division	Engineer's preliminary estimate	Contract price as modified by alterations	Final payment
Superintendent of Public Works.	....	Erie canal — Completing contract No. 47-A, east line of Wayne county to Lyons.	Western.	....	....	$917,880 30
State Engineer.	....	Erie canal — Completing contract No. 34, viaduct over Clyde river at Clyde.	Western.	....	....	3,799 20

2

TABLE OF CONTRACTS COMPLETED DURING THE FISCAL YEAR ENDED JUNE 30, 1919 — (Concluded)

Construction of Barge Canal Terminals

Chapter 746, Laws of 1911, and amendatory laws

CONTRACTOR	Date of contract	Character of work	Division	Engineer's preliminary estimate	Contract price as modified by alterations	Final payment
James P. Kelly	April 15, 1918	Terminal contract No. 8-P — Paving terminal at Schenectady	Eastern	$8,400 00	$8,400 00	$8,187 55
Patrick W. Mulderry	April 12, 1918	Terminal contract No. 10-P — Paving terminal at Fonda	Eastern	8,602 00	8,700 00	8,064 40
Anchor Post Iron Works	April 16, 1918	Terminal contract No. 12-F — Fence at Amsterdam terminal	Eastern			
Lut_per & Remick	Oct. 31, 1917	Terminal contract No. 15-M — Electrical equipment for the Utica terminal	Eastern	1,289 00	1,379 50	1,355 25
Walsh Construction Co., Inc	Nov. 4, 1915	Terminal contract No. 20 — Terminal basin with connecting basin to dike lake at Syracuse	Middle	30,681 20	36,967 50	37,069 72
Henry P. Burgard Co	May 6, 1918	Terminal contract No. 21-P — Paving part of terminal at Erie basin, Buffalo	Middle	665,875 00	549,878 26	*644,309 05
John E. Byron & Co	Oct. 30, 1916	Terminal contract No. 26 — Dredging and constructing pier at Rouses Point	Western	14,180 00	14,350 00	13,066 00
Patrick W. Mulderry	April 12, 1918	Terminal contract No. 27-P — Paving terminal at Frankfort	Eastern	51,200 00	55,678 50	*23,456 67
Barnaly & Ingersoll	Nov. 27, 1914	Terminal contract No. 29 — Harbor, dockwall and breakwaters on dike lake at Oswego	Eastern	4,100 09	4,446 00	3,938 45
Henry P. Burgard	Mar. 24, 1916	Terminal contract No. 30 — Dockwall and rough on east side of Geo river at Oswego	Middle	43,573 50	39,793 50	13,400 00
Lutpfer & Remick	Sept. 30, 1916	Terminal contract No. 31 — Terminal at Lyons	Middle	103,700 00	106,166 70	100,382 70
Guy B. Dickison	May 27, 1918	Terminal contract No. 33-P — Paving part of terminal pier at Oswego	Western	57,925 00	51,653 80	*56,521 09
Troy Public Works Co	Mar. 27, 1917	Terminal contract No. 36 — Terminal at Cohoes	Middle	11,010 00	11,730 00	11,329 00
Holler & Shepard	Aug. 26, 1915	Terminal contract No. 37 — Dockwall and harbor at Canajoharie	Eastern	61,000 00	57,600 00	*30,611 23
Geo. W. Rogers Co., Inc	June 8, 1917	Terminal contract No. 44 — Terminal at Mtt Haven	Eastern	33,832 00	32,272 00	31,436 63
M. H. Ripton	Oct. 19, 1916	Terminal contract No. 48 — Terminal on east side of Genesee weir at Rochester	Eastern	170,300 00	191,195 50	176,110 73
Kaufman & Garcey	July 27, 1916	Terminal contract No. 52 — Terminal at Pier 6, East 1 ier, New York city	Western	101,000 00	93,828 00	*94,766 00
I. J. Stander & Co., Inc	une 28, 1918	Terminal contract No. 56 — Repairing Pier 5, East river, New York city	Eastern	89,974 00	102,553 75	*132,148 66
W. F. Marteus	May 6, 1918	Terminal contract No. 59 — Road approach to terminal pier at Oswego	Eastern	20,400 00	27,159 60	28,904 68
			Middle	5,100 00	6,516 00	5,391 41

W. F. Martens	May 6, 1918	Terminal contract No. 60 — Railroad and crane tracks on terminal pier at Oswego.	Middle	8,365 00	9,690 00	9,119 00
H. W. Roberts & Co.	April 19, 1918	Terminal contract No. 63 — Railroad tracks and brick pavement on terminal at Utica.	Middle	9,590 00	7,672 00	7,632 13
Robert Wetherill, Receiver, American Pipe & Construction Co.	April 24, 1918	Terminal contract No. 64 — Railroad and crane tracks on terminal at Schenectady.	Eastern	9,000 00	10,021 30	10,227 44
Mohawk Dredge & Dock Co., Inc.	Dec. 18, 1916	Terminal contract No. 101 — Stiff-leg derricks on terminal sites at Albany, Whitehall, Little Falls, Rome, Lockport and Tonawanda.	Eastern, Middle and Western	21,890 90	31,790 90	36,611 92
The John F. Byers Machine Co.	Feb. 14, 1918	Terminal contract No. 106 — Fourteen two-ton tractor cranes for various terminals.	Eastern, Middle and Western	73,500 00	77,210 00	77,210 00
J. A. Laporte	Jan. 2, 1917	Terminal contract No. 201 — Terminal warehouses at Albany and Whitehall.	Eastern	59,300 00	71,432 30	*54,234 39
Savage Construction Co.	Feb. 14, 1918	Terminal contract No. 213 — Freight-house and four derricks at Syracuse.	Middle	28,200 00	26,997 00	28,346 40
Kennedy & Scullen Construction Co.	April 26, 1918	Terminal contract No. 214 — Freight-house and pavement at Amsterdam.	Eastern	16,478 00	16,323 00	14,709 00
Savage Construction Co.	July 9, 1918	Terminal contract No. 216 — Frame freight-house, Erie basin, Buffalo.	Western	10,000 00	9,899 00	10,116 00
James T. Young	Aug. 12, 1918	Terminal contract No. 220 — Addition to freight-house at Utica.	Middle	5,000 00	5,495 00	5,334 40
Kennedy & Scullen Construction Co.	Aug. 30, 1918	Terminal contract No. 221 — Freight-house at Herkimer and an extension to freight-house at Little Falls.	Eastern	6,100 00	6,122 00	5,914 10
J. A. Laporte	Aug. 23, 1918	Terminal contract No. 222 — Freight-house and roadway at Canajoharie.	Eastern	4,000 00	4,195 00	4,206 40

* Final payment made under chapter 585, Laws of 1918.
† Canceled August 8, 1917.

TABLE OF CONTRACTS PENDING JUNE 30, 1919

Special Work

CONTRACTOR	Date of contract	Character of work	Division	Act		Appropriation	Engineer's preliminary estimate	Contract price as modified by alterations	Value of work done to June 30, 1919
				Chap.	Year				
Scott Bros.	June 6, 1919	Constructing 8-foot pipe culvert across Barge canal at Rome	Middle	346	1918	....	$42,811 20	$46,731 20	*$000
E. Brown Baker	Nov. 23, 1918	Lake street bridge, Geneva	Middle	351	1918	....	55,811 00	70,400 00	*$200
Russell R. Ames	Jan. 30, 1919	Constructing a concrete culvert over Eighteen-Mile creek, Lockport	Western	626	1917	$12,500 00	10,905 00	11,236 00	*34000
Savage Construction Co.	May 31, 1919	Constructing a concrete culvert over Eighteen Mile creek, Lockport	Western	626	1917	12,500 00	10,905 00	11,070 00	0 00
Geo. L. Maltby	Mar. 23, 1916	Improvement of Chadakoin river, Chautauqua county	Western	{ 758 728	1913 1915 }	100,000 00	89,252 25	92,074 25	†8,570 00
Superin't of Public Works	....	Improvement of Chadakoin river, Chautauqua county	Western	{ 758 728	1913 1915 }	100,000 00	89,252 25	92,074 25	77,666 23

* Cancelled and relet to the Savage Construction Co.
† Cancelled, work being completed under the direction of Superintendent of Public Works.

Construction of the Barge Canal

Chapter 147, Laws of 1903; chapter 391, Laws of 1909; and amendatory laws

CONTRACTOR	Date of contract	Character of work	Division	Engineer's preliminary estimate	Contract price as modified by alterations	Value of work done to June 30, 1919
Walter S. Rae	April 15, 1918	Contract No. 117, Oswego canal — Bridge over lock No. 2, Fulton	Middle	$34,713 30	$36,513 80	$11,570 00
Peckham Construction Co., Inc	April 18, 1918	Contract No. 146, Erie canal — Movable dam at Herkimer	Eastern	81,726 20	93,769 40	51,800 00
Lathrop, Shea & Henwood Co	Sept. 10, 1917	Contract No. 147, Erie canal — Lift-bridge between Tonawanda and North Tonawanda	Western	227,032 80	234,260 40	78,170 00
Lathrop, Shea & Henwood Co	Sept. 5, 1917	Contract No. 148, Erie canal — Highway bridge at Leach street, Lyons	Western	65,810 60	66,986 20	56,440 00
Lathrop, Shea & Henwood Co	Oct. 30, 1917	Contract No. 164, Erie canal — Completing canal between Lyons and Newark and retaining dam at Macedon	Western	124,313 00	115,728 75	104,680 00
Walter S. Rae	Oct. 13, 1917	Contract No. 167, Oswego canal — Bascule bridge at Culvert street, Phoenix	Middle	26,653 60	29,689 30	18,460 00
Bronk & Kimmey	July 29, 1918	Contract No. 168, Champlain canal — Concrete-capped, timber guide-cribs near locks Nos. 3, 5 and 6	Eastern	63,505 60	85,727 40	67,240 00
I. M. Ludington's Sons, Inc	Nov. 9, 1917	Contract No. 179, Erie canal — Completing the canal at railroad crossings near Pittsford	Western	76,033 50	92,992 20	89,220 00
E. Brown Baker	Aug. 30, 1918	Contract No. 182, Oswego canal — Completing excavation in front of dockwall below lock No. 8, Oswego	Middle	28,215 00	30,267 00	26,660 00
Robert Wetherill, Receiver, American Pipe & Construction Co	June 24, 1918	Contract No. 185, Erie canal — Improving river channel below dams at Scotia and Rotterdam	Eastern	230,550 00	176,175 00	148,170 00
Scott Brothers	Aug. 20, 1918	Contract No. 187, Erie canal — Wash wall protection between New London and lock No. 22	Middle	17,525 00	22,530 00	15,470 00
E. Brown Baker	Aug. 7, 1918	Contract No. 188, Erie canal — Completing canal prism excavation at N. Y. C. R. R. bridge, Brewerton	Middle	30,000 00	35,400 00	30,260 00
Empire Engineering Co., Inc	Mar. 20, 1919	Contract No. 190, Erie canal — Completing canal from King's Bend to L. V. R. R. crossing at Rochester	Western	284,752 50	249,679 00	65,230 00

TABLE OF CONTRACTS PENDING JUNE 30, 1919 — (Continued)
Construction of the Barge Canal — (Continued)

CONTRACTOR	Date of contract	Character of work	Division	Engineer's preliminary estimate	Contract price as modified by alterations	Value of work done to June 30, 1919
Empire Engineering Co., Inc......	Jan. 14, 1919	Contract No. 191, Erie canal — Excavating canal channel in Genesee river near Elmwood avenue bridge...	Western.....	$189,850 00	$176,170 00	$7,360 00
Brown & Lowe Co..............	Jan. 22, 1919	Contract No. 192, Erie canal — Completing canal, east guard-lock to Genesee river, and work in Genesee valley park.............	Western.....	327,525 00	428,860 00	133,840 00
Stewart Bros..................	Feb. 27, 1919	Contract No. 197, Erie canal — Drilling holes in dam and lock near Rotterdam and making repairs.............	Eastern.....	27,170 00	27,780 00	8,190 00
Lupfer & Remick..............	Feb. 26, 1919	Contract No. 200, Erie canal — Driving sheet-piling, placing concrete lining, etc., between Rochester and Lockport.............	Western.....	257,992 50	180,248 50	11,340 00
I. M. Ludington's Sons, Inc.......	Mar. 13, 1919	Contract No. 201, Erie canal — Completing prism lining, Cartersville, and stream entrance Knapp's bridge...	Western.....	48,455 25	42,824 75	46,460 00
W. F. Martens................	Mar. 3, 1919	Contract Q, Cayuga and Seneca canal — Pile dolphins, Cayuga and Seneca lakes.............	Middle.....	5,225 00	5,092 00	5,090 00
Kennedy & Scullen Construction Company, Inc...........	Jan. 20, 1919	Contract T, Cayuga and Seneca canal — Extending core wall and other work at north end of dam No. 2, Seneca Falls.............	Middle.....	22,964 00	22,300 50	5,480 00
		Being Completed under chapter 585, Laws of 1918				
McArthur Bros. Co...........	Nov. 3, 1916	Contract No. 59, Erie canal — Constructing canal between contracts Nos. 21-A and 23-A at Genesee river, and Rochester harbor.............	Western.....	1,675,252 86	1,603,285 11	1,494,972 00
Combined Construction Co.......	April 19, 1917	Contract No. 138, Erie canal — Movable dam, etc., at Rochester.............	Western.....	302,700 30	321,115 12	471,363 00

Special Work Connected with Barge Canal Construction

CONTRACTOR	Date of contract	Character of work	Division	Engineer's preliminary estimate	Contract price as modified by alterations	Value of work done to June 30, 1919
Superintendent of Public Works..	....	Erie canal — Completing contract No. 63-A, west line of Wayne county to King's Bend...........	Western.....			$672,869 00

TABLE OF CONTRACTS PENDING JUNE 30, 1919 — (Concluded)

Construction of Barge Canal Terminals

Chapter 746, Laws of 1911, and amendatory laws

CONTRACTOR	Date of contract	Character of work	Division	Engineer's preliminary estimate	Contract price as modified by alterations	Value of work done to June 30, 1919
Barnally & Ingersoll	Feb. 15, 1915	Terminal contract No. 28 — Harbor, dockwall and breakwaters on Oneida lake at Cleveland	Middle	$34,575 00	$37,222 00	$35,120 00
I. J. Stander & Co.	Oct. 27, 1917	Terminal contract No. 38 — Constructing pier at West 53d street, North river, New York city	Eastern	250,000 00	265,550 39	214,390 00
Great Eastern Storage, Transfer & Wrecking Corp.	Nov. 29, 1918	Terminal contract No. 41 — Razing buildings and grading upper Troy terminal	Eastern	-1,500 00	-1,000 00	
Leonard Paving Co., Inc.	Nov. 13, 1918	Terminal contract No. 42 — Paving terminal at Long Island City	Eastern	54,600 00	53,579 00	1,370 00
Asphalt Construction Co.	May 13, 1919	Terminal contract No. 44-P — Paving at Mott Haven, Greenpoint and Gus bay terminals	Eastern	94,340 00	78,201 20	22,070 00
Sicilian Asphalt Paving Co.	June 13, 1919	Terminal contract No 52-P — Paving Pier 6, East river, New York city	Eastern	10,000 00	11,225 00	
Riverside Contracting Co.	Sept. 4, 1917	Terminal contract No. 55 — Terminal at Gus bay, Brooklyn	Eastern	513,000 00	507,620 40	260,080 00
Chas. Kiehm	Feb. 25, 1919	Terminal contract No. 57 — Part of an approach to terminal at Troy	Western	133,003 35	120,597 61	44,820 00
C. P. Boland & Co.	Dec. 2, 1918	Terminal contract No. 58 — Improving river terminal site at Troy	Eastern	16,600 00	19,400 00	7,260 00
Walsh Construction Co.	May 15, 1918	Terminal contract No. 61 — Railroad approach to pier 1, Erie basin, Buffalo	Western	9,720 00	11,650 00	490 00
Walsh Construction Co.	May 15, 1918	Terminal contract No. 62 — Railroad and crane take on pier 1, Erie basin, Buffalo	Western	8,470 00	11,400 00	9,420 00
Empire Engineering Co., Inc.	June 29, 1918	Terminal contract No. 66 — Riprap at Erie basin, Buffalo	Western	11,850 00	12,820 00	11,400 00
Walsh Construction Co.	July 3, 1918	Terminal contract No. 67 — Railroad approach to pier 2, Erie basin, Buffalo	Western	7,000 00	7,616 00	7,160 00
Walsh Construction Co.	July 3, 1918	Terminal contract No. 68 — Railroad track on pier 2, Erie basin, Buffalo	Western	6,820 00	7,445 00	4,630 00
Richard C. Bush	Feb. 27, 1919	Terminal contract No. 69 — Shore protection, Erie basin, Buffalo	Western	6,780 00	5,886 00	2,140 00
Geo. W. Chambers	April 9, 1919	Terminal contract No. 70 — Razing Bldgs and clearing State lands at Rochester	Western	-1,600 00	-4,267 00	
New Jersey Shipbuilding & Dredging Co.	May 13, 1919	Terminal contract No. 77 — Dredging at Piers 5 and 6, East river, at Greenpoint and at Long Island City	Eastern	40,325 00	42,895 00	6,770 00

Contractor	Date	Contract	Division			
Edward F. Terry Mfg. Co.	Jan. 21, 1919	Terminal contract No. 102 — Two 3-ton jib cranes, Pier 6, East river, New York city	Eastern	45,362 00	41,081 00	......
Brown Portable Conveying Machinery Co.	Oct. 3, 1917	Terminal contract No. 103-A — Two package-freight conveyors	Eastern	7,800 00	10,102 60	8,110 00
Lord Electric Co.	June 28, 1918	Terminal contract No. 105 — Installing electric wiring, lighting, power and battery-charging equipment, and auto-truck scales, Pier 6, East river, New York city	Eastern	17,742 40	16,000 50	......
J. Livingston Co.	Mar. 4, 1919	Terminal contract No. 107 — Electric lighting, power and battery-charging equipment, Erie basin, Buffalo	Western	35,025 00	28,238 50	8,370 00
General Electric Co.	Aug. 2, 1918	Terminal contract No. 109 — Electric capstans and trolley hoists at Pier 6, East river, and West 33 St. New York city, and electric capstan at Utica terminal	Middle and Eastern	19,000 00	18,536 00	......
General Electric Co.	June 9, 1919	Terminal contract No. 113 — Electric capstans and trolley hoists at various terminals		15,000 00	14,090 00	1,170 00
Electric Products Co.	June 20, 1919	Terminal contract No. 117 — Battery-charging and motor-generator sets for New York city terminals	Eastern	7,800 00	5,292 52	......
I. J. Stander & Co., Inc.	Jan. 11, 1918	Terminal contract No. 207 — Terminal freight-shed and head-house on Pier 6, East river, New York city	Eastern	133,500 00	128,250 01	......
Miller & Brady, Inc.	Mar. 22, 1918	Terminal contract No. 207-H — Heating system in freight-shed on Pier 6, East river, New York city	Eastern	3,250 00	2,352 00	113,390 00
Jarobo Bros., Inc.	April 16, 1918	Terminal contract No. 207-P — Plumbing and water-supply system in freight-shed on Pier 6, East river, New York city	Eastern	6,000 00	6,650 00	1,170 00
Felton Construction Corp.	Nov. 14, 1918	Terminal contract No. 212 — Freight-house, Pier 1, Erie basin, Buffalo	Western	175,000 00	181,669 00	......
A. E. Norton, Inc.	Oct. 23, 1918	Terminal contract No. 217 — Repairing bulkhead and constructing freight-house at Long Island City	Eastern	59,850 00	74,806 50	5,190 00
P. Altman Plumbing Co.	May 29, 1919	Terminal contract No. 217-P — Plumbing and water-supply systems, Long Island City terminal	Eastern	4,000 00	3,765 00	15,510 00
Donnell-Zane Co., Inc.	May 13, 1919	Terminal contract No. 218 — Freight-shed at West 53d street, New York city	Eastern	53,969 15	46,549 20	48,860 00
Post & McCord	Mar. 20, 1919	Terminal contract No. 223 — Freight-shed at Greenpoint terminal	Eastern	99,710 00	75,718 94	1,390 00
J. A. Laporte	April 28, 1919	Terminal contract No. 226 — Freight-house at river terminal, Oswego	Middle	6,000 00	5,199 00	5,030 00
J. A. Laporte	April 28, 1919	Terminal contract No. 227 — Freight-house at Gowanus bay	Eastern	9,760 00	9,189 00	8,000 00

Being Completed under Chapter 585, Laws of 1918

Contractor	Date	Contract	Division			
McHarg-Barton Co.	Nov. 24, 1916	Terminal contract No. 19 — Terminal at Greenpoint	Eastern	39,500 00	211,513 00	282,834 00
Empire Engineering Co., Inc.	Jan. 12, 1914	Terminal contract No. 21 — Terminal at Erie basin, Buffalo	Western	925 00	797,772 30	897,875 00
Walsh Construction Co.	Oct. 27, 1916	Terminal contract No. 53 — Terminal at Ohio basin, Buffalo	Western	571,800 00	597,984 00	387,513 00

SUMMARY OF CONSTRUCTION WORK — BARGE CANAL AND TERMINALS

Value of work done under Barge canal and terminal contracts, summarized by years and canals

BARGE CANAL

YEAR*	ERIE CANAL			Champlain canal	Oswego canal	Cayuga and Seneca canal	Barge canal terminals
	Eastern Division	Middle Division	Western Division				
1905	$36,640		$59,190	$71,620			
1906	140,860	$52,570	197,840	216,950	$2,220		
1907	553,980	228,820	192,580	658,580	69,010		
1908	1,557,774	580,367	416,290	1,426,159	173,030		
1909	2,325,760	1,530,410	2,012,119	1,281,610	324,060		
1910	1,994,920	1,394,542	3,279,387	1,361,668	529,890		
1911	2,756,161	2,362,223	5,285,424	2,342,781	1,075,556	$432,050	
1912	2,128,787	1,802,231	6,595,375	1,070,702	1,117,259	517,450	
1913	2,084,109	1,682,835	5,380,484	1,227,017	1,427,068	865,810	$569,410
1914	2,593,664	2,079,504	2,334,353	2,038,534	1,059,637	1,231,539	1,119,600
1915	2,996,147	1,146,418	476,377	1,240,491	552,925	1,365,450	1,009,173
1916	1,312,713	250,385	412,228	406,928	177,856	412,492	462,935
1917	1,277,160	385,240	1,179,508	573,694	56,340	167,091	970,633
1918	412,967	385,468	2,673,153	254,523	60,519	142,931	1,038,868
1919	323,488	100,607	3,450,346	216,266	121,037	178,215	1,688,852
Totals	$22,505,130	$14,176,614	$33,944,652	$14,388,553	$6,746,397	$5,313,028	$6,839,471

Extra Work Orders Paid

YEAR	Eastern Division	Middle Division	Western Division	Champlain canal	Oswego canal	Cayuga and Seneca canal	Barge canal terminals
1906	$1,316						
1907	378	$1,257					
1908	13,020	50	$300	$59,738			
1909	37,141	329	10,464	23,854	$1,198		
1910	12,806	12,380	10,692	16	761		
1911	22,102	3,353	175,606	1,001	2,135		
1912	34,957	2,899	94,935	14,524	1,703		
1913	76,860	11,101	203,577	34,810	8,416	$2,066	
1914	12,969	17,878	239,744	30,718	15,719	9,534	$2,033

1915	19,731	11,434	20,954	11,094	4,130	2,926	10,437
1916	8,728	10,355	436	290	276	66,709	14,134
1917	12,699	3,714	8,893	1,844	10,217	8,845	9,995
1918	123,585	13,163	142,006	11,376	739	9,796	24,725
1919	45,967	1,831	130,285	3,595	....	221	15,117
Totals	$422,349	$89,753	$1,037,881	$192,860	$45,293	$100,096	$76,441

* The years 1905 to 1915, inclusive, are twelve-month periods, ended September 30; 1916 is a nine-month period, ended June 30; 1917, 1918 and 1919 are twelve-month periods, ended June 30.

NOTE.— This table includes work done under the supervision of this Department, excepting highways which were relocated or rebuilt; also the following items: Contract No. 20-D, work done by the Superintendent of Public Works, $3,400; contract No. 20-D, special agreement, $64,816; contract No. 21-A, special agreement for erecting steel and machinery on guard-lock, $4,874; contract No. 22, special agreement, $12,447; contract No. 25, special agreement, $6,029; contract No. 47-A, work done by Superintendent of Public Works, $617,880; culvert No. 30, at crossing of Irondequoit creek, $372,549; shelter at Delta dam, $2,234; and contract No. 63-A, work done by the Superintendent of Public Works, $672,869.

REPORT

OF THE

DIVISION ENGINEER

OF THE

EASTERN DIVISION

For the Fiscal Year Ended June 30, 1919

[45]

EASTERN DIVISION

STATE OF NEW YORK

DEPARTMENT OF STATE ENGINEER AND SURVEYOR

EASTERN DIVISION

ALBANY, N. Y., *July* 1, 1919.

Hon. FRANK M. WILLIAMS, *State Engineer and Surveyor.*
Albany, N. Y.:

Sir.— I have the honor of submitting the following report of the work of the Eastern Division for the fiscal year ended June 30, 1919.

Mr. George D. Williams, Division Engineer, has been in military service during this period. For the period from July 1 to December 31, 1918, Mr. L. C. Hulburd, Senior Assistant Engineer, was in charge, and from January 1, 1919, to date the undersigned has had the honor of directing the work of the Division.

BARGE CANAL AND TERMINALS

Construction work on the Barge canal on this Division has been confined to such contracts as would perfect the canal and make its operation more satisfactory, and also to such work as would correct some conditions that have developed since the canal was put into use. In doing this work the canal has been made safer in operation. For example, a new movable dam has been under construction at Herkimer to replace a needle dam, and at three places in the Champlain canal where cross currents existed in the channel, guide cribs have been placed, to aid boats in keeping the channel.

A considerable portion of the work has been such as would naturally be necessary to maintain so large a plant as is the Barge canal. This has consisted in strengthening some few structures, in removing bars, such as must annually be found in the channel, particularly at points where live streams carry into the canal large amounts of sand, gravel and other material, in adding to and repairing riprap and wash walls and in giving close attention to repairs to structures and operating machinery. The work of maintenance is under the direction of the Superintendent of Public Works, but in this work this Department has had an active part by furnishing the engineers to care for the engineering features of the work and frequently to advise as to extent and methods. Hearty and sympathetic coöperation between the two departments on this Division has resulted in securing most satisfactory results in the maintenance work. It should be noted that with an apparently considerable increase in traffic through the canals on this Division there has been very little complaint on the part of boatmen of difficulties or obstacles to navigating their boats and fleets.

With the completion of new construction on the canals on this Division the problems will continue to be those already met in maintaining these canals. The plan of coöperation between the Department of Public Works and this Department, already referred to, can be productive of the greatest efficiency if properly organized engineering parties are maintained to keep in touch with the conditions of the canals and to be ready to perform the work of aiding in the correction of any defects that may develop from either ordinary or unusual conditions. In this connection I would respectfully suggest that provision should soon be made for securing some well equipped sweep-boats. With these, obstructions in the channel can be more readily discovered and danger of damages to boats be materially lessened.

Work on the terminals for the Barge canal has been confined on this Division mostly to that on the terminals at New York, although work has also been done during the year to terminals at Troy, Amsterdam, Little Falls, Herkimer and Canajoharie. Warehouses of steel and tile construction have been built at Albany and Whitehall. The work at New York has been quite extensive, as will be noted in the report of the Senior Assistant Engineer in charge of the work. A feature of all work at the terminals has been the installation of various types of freight-handling equipment.

FLEET OF MODERN STEEL CANAL BARGES

Most of the boats built especially for Barge canal traffic have been of about the size shown here — 150 by 21½ feet, of some 650 tons capacity. Although several scores have been built, the number is lamentably small in view of the possible amount of traffic, if shipping were available.

A BOAT TYPICAL OF SEVERAL ON THE CANAL

A few large industrial corporations have built their own boats and are operating them on the Barge canal. This boat and several more like it belong to a well-known oil company.

Navy Coal Barges on the Canal

The Barge canal is very frequently used for transporting boats built at inland shipyards and intended for ocean or harbor service.

BARGE CANAL TERMINAL AT ALBANY

The warehouse, a large and substantial concrete structure, was recently completed.

LOWER TERMINAL AT TROY

View when a large number of canal-boats were lying at the dock.

A LOCKAGE OF OLD-SIZED CANAL-BOATS

Six of these boats and a small tug may be locked through at a single lockage. Not many old boats were available when the new canal was completed and their number is steadily diminishing. The new boats are of larger size.

In general the work has progressed very satisfactorily, except in a few cases. Existing conditions in the labor and materials markets have seriously handicapped some of the work. It should be said, however, that most of the contractors have made all reasonable efforts to progress their work and that such delays as may have occurred were largely beyond their control.

MISCELLANEOUS WORK

In addition to construction and maintenance work on the canals and terminals this Division has made many investigations of claims for alleged damages; has investigated and reported on applications for reconveyances of lands; has prepared maps for lands no longer required for canal purposes, and has prepared maps for the appropriations of additional lands where construction work made such appropriations necessary.

SPECIAL APPROPRIATIONS

In addition to the canal improvement this Department has been employed in preparing plans and supervising contracts for work provided under special acts of the Legislature and in making miscellaneous surveys, maps, plans, etc., for other State departments. A brief outline of such activities follows:

SURVEYS

Blue Line Surveys
(Chapter 119, Laws of 1910, and amendatory laws)

The work of surveying and mapping old canal lands has been continued. Surveys for the Champlain canal are now complete except for the Glens Falls feeder and that portion between Waterford and the junction with the Erie at Colonie. These sections are to be retained for use as parts of the new canal system.

On the Erie canal surveys are now complete from the easterly end of the canal to Fort Hunter, except the portion between lock No. 1 and the junction with the Champlain canal at Colonie. Surveys from Fultonville to Randall, from Sprakers to Mindenville and through the city of Little Falls are also complete. Surveys are under way between Fort Hunter and Fultonville, Randall and Sprakers, and Mindenville and Little Falls. The plotting of the maps is well up with the survey work. It is expected that this work, and that between Little Falls and Mohawk will be completed by October 31, 1919.

Terminals Survey
(Chapter 555, Laws of 1918)

Under this act a survey was made of the site at Poughkeepsie, on which it is proposed to locate a Barge canal terminal.

Delaware–Schoharie County Line Survey
(Chapter 559, Laws of 1918)

This act provided for the survey of that portion of the boundary line between the counties of Delaware and Schoharie that is between Schoharie creek and Lake Utsayantha. This line was surveyed and properly monumented during the summer of 1918, as reported in a special report.

Boundary Line Survey between Towns of Warrensburg and Luzerne, Warren County, and between Warren and Saratoga Counties
(Chapter 561, Laws of 1918)

Under this act the line between the towns of Luzerne and Warrensburg in Warren county and the line between the counties of Saratoga and Warren was to be surveyed and monumented. The appropriation for this work was not sufficient for making the survey of the whole line, but the portion which was most desired to be located, that between Fort George at Lake George and the Hudson, was completed and monumented. A special report of this survey was also made.

Greene–Ulster County Line Survey
(Chapter 562, Laws of 1918; chapter 600, Laws of 1919)

These acts provided for the location of a portion of the boundary line between the counties of Greene and Ulster. This work was started in 1918, but, in order to determine the proper location of that portion of the line specified in the acts, it became necessary to survey a much longer line and the work had to be continued in 1919. This survey has now been completed and the portion defined in the acts has been properly monumented, as will be noted in a special report.

Miscellaneous

A survey was made under chapter 317, Laws of 1917, of a proposed route for the Jamaica Bay–Peconic Bay canal through Far Rockaway.

An extensive survey for a proposed exterior belt line was made for the New York–New Jersey Port and Harbor Development Commission.

Under various acts of the Legislature surveys have been made by parties of this Division at the new prison, Wingdale, at Craig Colony hospital, Sonyea, at Kings Park hospital, L. I., at St. Lawrence hospital, Ogdensburg, and for a proposed improvement of Mills river, L. I.

Surveys of lands under water were made at New York city and at several points in its immediate vicinity, also at Troy, Poughkeepsie and Tarrytown.

State Boundary

The boundary line between the states of New York and New Jersey was examined and the condition of the various monuments on the line noted.

Stream Gages

For the use of the Court of Claims numerous stream gages have been set at various points. One of special value is the gage placed on the Nine-Mile creek feeder.

CONSTRUCTION WORK

Removal of High Street Bridge, Cohoes
(Chapter 181, Laws of 1917; chapter 151, Laws of 1918)

Under these acts the bridge at High street, Cohoes, which crossed the abandoned portion of the Erie canal at this point, was removed, the canal filled in and the approaches properly regraded.

Roadway at Rockaway Point, L. I.
(Chapter 130, Laws of 1917)

A concrete road was built under this act at Rockaway Point.

Repairing Sea-walls, Orient–East Marion, L. I.
(Chapter 428, Laws of 1918)

Under this act repairs were made to the sea-walls between the villages of Orient and East Marion, L. I.

Repairing Landing Piers, Hoffman and Swinburne Islands

Repairs were made to the landing piers at Hoffman island and at Swinburne island.

Schenectady–Scotia Bridge
(Chapter 735, Laws of 1917; chapter 624, Laws of 1919)

Under these acts parties have located on the ground, the center line of the bridge, have surveyed for the approaches at the Schenectady and Scotia ends and for the lands to be appropriated and have made extensive soundings and test pits.

In conclusion I would like to take this opportunity of expressing my appreciation for the assistance and advice which you and your deputies have given me, and also to state that I appreciate the hearty response made by the men on this Division to any tasks assigned to them. I believe that they have all been faithful and efficient.

Detail reports by those who have been in charge of the residencies into which the Division is divided, together with tabulations showing financial statements and disbursements are appended. In these reports will be found descriptions of the various pieces of work done during the year.

Respectfully submitted,

RUSSELL S. GREENMAN,
Senior Assistant Engineer in Charge.

APPENDED REPORTS—EASTERN DIVISION

ERIE CANAL, RESIDENCY No. 1

Assistant Engineer in charge R. D. Hayes reports:

This residency extends from Albany to the site of the old lower Mohawk aqueduct. The work has been directed from the Mechanicville office.

During the year several claims were investigated and reports made to the Division office.

Reports on contract No. 91 and terminal contracts Nos. 36, 41 and 58 and a part of terminal contracts Nos. 101, 106 and 201 and on the removal of High street bridge, Cohoes, follow.

Contract No. 91

This contract is for building and equipping a hydro-electric power-plant on the Erie canal near the east end of Crescent dam. It was awarded to Welles-Boughton & Co., being signed on January 5, 1911. It was assigned to the Hollington Company, this assignment being approved by the Superintendent of Public Works July 31, 1911. Construction work began April 3, 1911. The engineer's preliminary estimate was $44,600.00, the contractor's bid, $42,940.50. The contract price as modified by alteration No. 1 is $44,985.50. The work was accepted December 27, 1918, and the final account, amounting to $40,458.32,* was approved by the Canal Board December 27, 1918.

No work was done during the year.

Terminal Contract No. 101

This contract is for furnishing and installing steel stiff-leg derricks on terminal sites at Albany, Whitehall, Little Falls, Rome, Lockport and Tonawanda. It was awarded to E. Brown Baker, being signed on December 18, 1916. On February 21, 1917, it was assigned to the Mohawk Dredge and Dock Co., Inc., and this assignment was approved by the Superintendent of Public Works March 26, 1917. The engineer's preliminary estimate was $21,890.90,

---

* The sum of $3,292 has been deducted to cover actual cost to the State of providing for a new governor equipment under contract No. 91-A.

the contractor's bid, $31,790.90. Excess metal to the value of
$6,510 has been authorized by the Canal Board. The work was
accepted December 4, 1918, and the final account, amounting to
$36,611.92, was approved by the Canal Board December 27, 1918.
The amount paid on extra work orders during the year was $2,966.-
03, total to date, the same.

One of these derricks has been installed at Albany, the final
account for which amounted to $6,124.56.

W. L. Caler, Assistant Engineer, was in charge.

Terminal Contract No. 201 — Albany

This contract is for constructing terminal warehouses at Albany
and Whitehall. It was awarded to J. A. Laporte, being signed on
January 2, 1917. The engineer's preliminary estimate was $59,-
300.00, the contractor's bid, $65,174.85. The contract price as
modified by alteration No. 1 is $71,432.30. Construction work
began at Albany on April 3, 1917. The engineer's preliminary
estimate for this warehouse was $36,500.00, the contractor's bid,
$40,106.50.

W. L. Caler, Assistant Engineer, is in charge.

Under authority of chapter 585, Laws of 1918, this contract was
canceled by the Canal Board on July 17, 1918. The cancelation
became effective August 31, 1918, on approval of the Canal Board,
the contractor having filed a stipulation of his compliance with the
terms of the law. The actual cost of the work from April 7, 1917,
to August 31, 1918, was $42,790.16 and payment of balance due on
this amount was authorized by the Canal Board on February 26,
1919. No work was done prior to April 7, 1917.

The contractor has completed the work and the total payments to
date, including extra work orders, are $54,234.39.

At the Albany warehouse during the year the paving of the
depressed roadway and around the warehouse was completed. The
railroad siding was built. The side walls were completed and are
being stuccoed. The roof was placed by October. The chimney
was finished. The partitions were put in, the metal ceiling was
built over the second floor and the stairs and railing around the
stair well were placed. The skylights, the entrance doors, windows
and the doors in both the warehouse and the office were placed, and

the copper roof was put on the marquise. Throughout the building, in both warehouse and office portions, the plumbing was completed except fixtures and also the wiring except the fixtures.

Terminal Contract No. 106

This contract is for furnishing fourteen two-ton steam tractor cranes for terminals. It was awarded to John F. Byers Machine Co., being signed on February 14, 1918. The engineer's preliminary estimate was $73,500.00, or $5,250 per crane, the contractor's bid, $73,710.00, or $5,265 per crane. The contract price as modified by alteration No. 1 is $77,210.00, or $5,515 per crane. The value of work done during the year is $37,080. The work was accepted September 24, 1918, and the final account, amounting to $77,210.00, was approved by the Canal Board October 9, 1918.

C. A. Curtis, Assistant Engineer, was in charge.

One of these cranes was delivered at the Troy lower terminal in June, 1918. The official tests were made during the past year. The final account for the work at Troy amounted to $5,515.00.

Terminal Contract No. 58 — Troy, Lower Terminal

This contract is for improving the lower terminal site at Troy. It was awarded to C. P. Boland & Co., signed on December 2, 1918. Construction work began May 20, 1919. The engineer's preliminary estimate was $16,600.00, the contractor's bid, $19,400.00. The value of work done during the year is $7,260, total done to date, the same.

W. L. Caler, Assistant Engineer, is in charge.

During the year the excavation for the warehouse extension piers, the macadam and the brick pavements was made, the four-inch tile drain laid, the six-inch concrete edging completed, all track in stone ballast laid, and the six-inch base for the brick pavement and the bottom course for the macadam roadway placed. The warehouse extension has been practically finished.

Terminal Contract No. 41 — Troy, Upper Terminal

This contract is for razing the buildings and grading the upper terminal site at Troy. It was awarded to Great Eastern Storage, Transfer & Wrecking Corporation, being signed on November 29,

1918. Construction work began January 23, 1919. The engineer's preliminary estimate was that the contractor should pay the State $1,500.00, the contractor bid that he would pay the State,$1,000.

W. L. Caler, Assistant Engineer, is in charge.

All buildings, with the exception of Nos. 6 and 16, have been razed. Material obtained from outside sources was used to fill in forebay under building No. 15. Material is now being placed along the street to form the slope at the north end of the site. No excavation has been made at the site.

Terminal Contract No. 36 — Cohoes

This contract is for constructing a terminal at Cohoes. It was awarded to the Troy Public Works Co., being signed on March 27, 1917. The engineer's preliminary estimate was $61,000.00, the contractor's bid, $57,600.00.

F. W. Harris, Assistant Engineer, was in charge.

Under authority of chapter 585, Laws of 1918, this contract was canceled by the Canal Board on July 9, 1918. The cancelation became effective October 9, 1918, on approval of the Canal Board, the contractor having filed a stipulation of his compliance with the terms of the law. The actual cost of the work from April 7, 1917, to October 9, 1918, was $30,611.23 and payment of balance due on this amount was authorized by the Canal Board on December 27, 1918. No work done prior to April 7, 1917.

This contract has been completed and the total payments, including extra work orders, was $30,611.23.

The work originally contemplated under this contract is to be completed under a new contract, No. 36-A.

Removal of High Street Bridge, Cohoes
(Chapter 181, Laws of 1917; chapter 151, Laws of 1918)

This contract is for the elimination of the High street bridge at Cohoes and the substitution of earthen embankments, together with the widening of Sandusky street and other incidental work. It was awarded to Michael Fitzgerald, being signed on September 9, 1918. Construction work began about September 10, 1918. The engineer's preliminary estimate was $13,000.00, the contractor's bid, $11,511.50.

C. A. Curtis, Assistant Engineer, is in charge.

Embankment was made by wheel scraper outfits, two of which were used. Work was continued until December, when a steam air-compressor plant was set up and the various concrete and masonry walls were demolished. Embankment work was resumed in the spring of 1919 and practically finished by June 30, 1919. In cutting down the old bridge an acetylene torch was used. Part of this structure was sold to the city and the balance as junk.

ERIE CANAL, MOHAWK RIVER RESIDENCY

Assistant Engineer M. E. James reports:

When, on June 1, 1919, the residency office at Little Falls, on residency No. 4, was closed, the Mohawk River residency was extended on the west from Mindenville lock, No. 16, to the Herkimer–Oneida county line. The Mohawk River residency, in addition to the extent originally assigned to it, now has jurisdiction over all work formerly under Erie residencies Nos. 2, 3 and 4. It extends now from Crescent dam to the Herkimer–Oneida county line, a distance of 96 miles.

E. A. Lamb, Senior Assistant Engineer, was in charge of residency No. 4 during the fiscal year up to June 1, 1919, and his report is incorporated herewith.

The engineering forces on this residency, in addition to looking after construction work, have devoted a portion of their time to the engineering work necessary in connection with the locating of buoys for navigation, in sweeping the navigation channel, in investigating and reporting on claims and complaints arising in connection with the construction and operation of the Barge canal and in making investigations and reports in connection with the release of appropriated lands which, in whole or in part, are no longer needed for the use of the canals, and also investigations and reports in connection with repairs and maintenance of the canal.

Reports on contracts Nos. 20-D (extra work order), 29-A, 122-A, 146, 155 (extra work order), 180, 181, 185 and 197, terminal contracts Nos. 8-P, 10-P, 12-F, 27-P, 37, 64, 214, 221 and 222, also the portions of terminal contracts Nos. 101, 103-A and 106 within this residency, follow:

Contract No. 20-D

This contract, which was for dredging a channel in the Mohawk river from Yosts to Rexford, was completed and final estimate approved prior to the time of this report, but an extra work order under this contract had not been finished. The amount paid on extra work order during the year is $44,998.94, total to date, $197,228.40.

William M. Griffith, Junior Assistant Engineer, was in charge of this work.

An extra work order dated January 15, 1917, provides for the construction of a coffer-dam around the north span of dam No. 5, at Rotterdam, and making repairs to this dam. The final account, amounting to $132,003.20, was approved by the Canal Board April 2, 1919. Since one of the largest items in this amount was the cost of the steel sheet-piling purchased but subsequently used elsewhere, the cost of the work would be materially reduced by making a reasonable credit for this item.

Under this work order the American Pipe and Construction Co. drilled in the north span of the dam 62 eight-inch holes with a well-drill and 45 three-inch holes with the steam-drill, for the purpose of exploring the condition of the foundation beneath the sill and for filling such voids as might be found. Through these holes a total of more than 200 yards of concrete, grout and gravel were deposited under the sill of the dam. During August four clusters of piles were driven at the lower entrance to the lock, No. 9, as an aid to navigation. An open hole was dug back of the land wall of the lock to investigate the cause of a settlement in the paving and any possible leakage or current of water under the land wall, but nothing was discovered.

During September and October the plant was removed from inside the coffer-dam, the dam gates were lowered and during November the removal of the coffer-dam above the north span was begun by a dipper-dredge. The steel sheet-piling and timber were stored back of the lock wall. On December 6 the closing of the river by ice necessitated the removal of the dipper-dredge to a winter harbor and the balance of the coffer-dam was removed by hand after the gates of the dam were raised. All work was completed and the contractor's plant removed from the site of the work during the month of January, 1919.

Contract No. 155

This contract, which was for furnishing and installing seven hoists for the operation of the bulkhead gates in the north end of the Vischer Ferry dam, was completed and final estimate approved prior to the time of this report, but an extra work order under this contract had not been finished. The amount paid on extra work order during the year is $930.36, total to date, the same.

C. B. Tebo, Engineering Assistant, was in charge of this work.

An extra work order dated March 25, 1918, provides for repairing hoists damaged at the time of the spring flood. The final account, amounting to $930.36, was approved by the Canal Board May 21, 1919.

The anchorages of the hoists had been repaired, the gears delivered and all but two, which were found to be defective upon delivery, had been installed prior to July 1, 1918. During September and December, 1918, these gears were repaired and installed, completing the work called for under this work order.

Contract No. 180

This contract was for removing a portion of the aqueduct at Rexford and completing the adjacent canal prism excavation. It was awarded to Dunbar & Sullivan Dredging Co., being signed on March 15, 1918. Construction work began April 17, 1918. The engineer's preliminary estimate was $17,840.00, the contractor's bid, $15,958.00. The work was accepted December 18, 1918, and the final account, amounting to $16,153.54, was approved by the Canal Board December 18, 1918.

M. J. Quinn, Junior Assistant Engineer, was in charge.

This work was completed prior to July 1, 1918, the final estimate being prepared during the current year.

Contract No. 185

This contract is for improving the river channel below dams at Scotia and Rotterdam. It was awarded to Robert Wetherill, Receiver, American Pipe and Construction Co., being signed on June 24, 1918. Construction work began July 22, 1918. The engineer's preliminary estimate was $230,550.00, the contractor's bid, $154,395.00. The contract price as modified by alteration

No. 1 is $176,175.00. The value of work done during the year is $148,170.00, total done to date, the same.

M. J. Quinn, Junior Assistant Engineer, was in charge until January 1, 1919. Since that time A. P. Mussi, Assistant Engineer, has been in charge.

Alteration No. 1, approved by the Canal Board May 7, 1919, provides for widening the river channel on the south side between Stas. 1625 and 1648 below the dam at Rotterdam. It increases the contract price by $21,780.00.

The hydraulic dredge *Mohawk,* belonging to the American Pipe and Construction Co., arrived at Schenectady on July 6, 1918, was set up and began excavating in the main channel below dam No. 4 at Scotia on July 22, spoiling the excavated material in the old Erie canal. After completing the excavation of refill in the main channel south of the island, the dredge moved into the north channel below the dam and started to deepen and widen the channel around the north side of the island. This excavated material was spoiled in the bed of the old canal, on the island and on the Cramer parcel on the north bank below the dam. The excavation below dam No. 4 was completed about the middle of October and the dredge moved to below dam No. 5 at Rotterdam, where excavation was started October 17, the excavated material being placed in the spoil-bank along the south bank below the dam. The closing of the river by ice on December 6 compelled the removal of the dredge to winter harbor before completing the excavation at dam No. 5. Before the beginning of work in the spring of 1919 alteration No. 1 was executed; this extends the limit of the excavation downstream in order to widen the channel below dam No. 5 and remove a portion of the gravel bar which contracts the channel at that point.

A derrick was set up on the island below dam No. 4 in October, 1918, and first-class riprap stone was delivered. The placing of this riprap on the upper point of the island was completed during November and the plant removed.

To July 1, 1919, the monthly estimates on this contract amounted to 288,851 cu. yds. of excavation and 1,155 cu. yds. of first-class riprap.

The contract work was completed June 20, 1919, and the final estimate is being prepared.

Contract No. 197

This contract is for drilling holes in the sill of the dam and the toe of the river wall of the lock near Rotterdam and making necessary repairs. It was awarded to J. W. Holler, being signed on January 27, 1919. On January 28, 1919, it was assigned to Stewart Brothers and this assignment was approved by the Superintendent of Public Works February 25, 1919. Construction work began on January 29, 1919. The engineer's preliminary estimate was $27,170.00, the contractor's bid, $27,780.00. The contract price as modified by alteration No. 1 is $27,780.00. The value of work done during the year is $8,190.00, total done to date, $8,190.00.

William M. Griffith, Junior Assistant Engineer, is in charge.

Alteration No. 1, approved by the Canal Board August 6, 1919, eliminates the item "diverting dams" and provides a new item for performing the work in a different manner. The contract price remains the same.

On January 29 the plant was delivered and during February holes were drilled through the foundation of the river wall of the lock, disclosing voids. Holes were drilled also through the sill of the south span of the dam, but no voids were found. Because of difficulty with the boilers used in running the steam-drills, work that could have been performed on the ice was delayed until after the ice went out on March 1. In March holes were drilled along the river wall of the lock, disclosing voids of from four to eight feet in depth. Some concrete was deposited through these holes, but work was delayed somewhat by the condition of the river. During the latter half of March, after the Department of Public Works had lowered the gates in the middle span of the dam to allow the contractor to float a scow, it was attempted to drill holes in the middle span with tripod drills, but with poor success. A well-drill replaced the steam-drill and after April 9 the drilling of exploratory holes progressed rapidly. A void of five feet in depth was found under the sill on the south side of the north pier, but no indication of voids was found toward the south pier.

tion work began in April, 1916. The engineer's preliminary
estimate was $162,005.00, the contractor's bid, $185,106.50. The
contract price as modified by alterations Nos. 1 to 5, inclusive, is
$318,659.70. Excess quantities to the value of $2,039.00 have
been authorized by the Canal Board. The amount paid on extra
work orders to date is $4,944.60.

C. W. Wilbur, Junior Assistant Engineer, was in charge.

Under authority of chapter 585, Laws of 1918, this contract
was canceled by the Canal Board on July 9, 1918. The cancelation became effective November 13, 1918, on approval of the
Canal Board, the contractor having filed a stipulation of his
compliance with the terms of the law. The actual cost of the
work from April 7, 1917, to November 13, 1918, was $219,885.17
and payment of balance due on this amount was authorized by the
Canal Board on May 21, 1919. The final account for work done
prior to April 7, 1917, amounting to $151,613.03, was author-
ized by the Canal Board on May 7, 1919.

This contract has been completed and the total payments, including extra work orders, was $376,442.80.

During the year approaches to Schuyler bridge were completed,
concrete at Harbor bridge placed, wash wall and embankments
trimmed and refill at various points, notably at Knapp's brook,
removed. Contract work was practically completed by October
10, 1918.

Terminal Contract No. 8-P — Schenectady

This contract is for paving the terminal site at Schenectady.
It was awarded to James P. Kelley, being signed on April 15,
1918. Construction work began May 6, 1918. The engineer's
preliminary estimate was $8,400.00, the contractor's bid,
$8,400.00. Excess excavation to the value of $104.00 has been
authorized by the Canal Board. The value of work done during
the year is $7,187.55. The work was accepted December 18,
1918, and the final account, amounting to $8,187.55, was approved
by the Canal Board January 22, 1919. The amount paid on
extra work orders during the year is $265.14, total to date, the
same.

M. J. Quinn, Junior Assistant Engineer, is in charge.

TYPES OF BOATS USING THE CANAL

At the right, a concrete car-float, being shipped from the Great Lakes for use in New York harbor. At the left, a luxurious private yacht. In the distance, a fleet of freight-boats.

SHIPYARD AT WATERVLIET

This yard is constantly engaged in building boats for Barge canal use. One of these boats is shown here.

AN UNUSUAL CANAL-BOAT

This boat and its companion, the *Idlewild*, although old and never before engaged in canal traffic, were impressed into service for sending a cargo of molasses through the canal.

ANOTHER UNUSUAL CANAL-BOAT

The shortage of available shipping led to the use of this boat and its companion, the *Louisa*, as canal boats. There is an abundance of canal freight, but, until boats are provided, its transportation is deplorably limited.

An extra work order dated November 26, 1918 provides for constructing a manhole and a drainage ditch. The final account, amounting to $265.14, was approved by the Canal Board January 15, 1919.

The construction of the brick pavement under this contract could not be progressed until the track work under terminal contract No. 64 was completed. Hence no work was done during July or August, 1918. During September the brick pavement was constructed and during October and November the macadam pavement at the Fuller street entrance was built and that around the freight-house repaired, which completed the work under the original contract. During December the drainage ditch back of the brick pavement and the manhole over Cowhorn creek were constructed under the extra work order.

All work was completed by December 13, 1918.

Terminal Contract No. 64 — Schenectady

This contract is for constructing railroad and crane tracks on terminal at Schenectady. It was awarded to Robert Wetherill, Receiver, American Pipe & Construction Co., being signed on April 24, 1918. Construction work began May 1, 1918. The engineer's preliminary estimate was $9,000.00, the contractor's bid, $10,021.30. The value of work done during the year is $8,077.44. The work was accepted November 13, 1918, and the final account, amounting to $10,227.44, was approved by the Canal Board December 18, 1918. The amount paid on extra work orders during the year is $795.00, total to date, the same.

M. J. Quinn, Junior Assistant Engineer, was in charge.

An extra work order dated June 4, 1918, provides for making alterations and extensions to the electrical equipment at the Schenectady terminal freight-house. The final account, amounting to $795.00, was approved by the Canal Board December 4, 1918.

During July, 1918, the concrete ramp south of the freight- house was removed and the gap in the dockwall was filled with concrete.

One railroad track was laid and material for the other track was delivered. The rewiring of the freight-house under the extra work order dated June 4 was practically completed. During August the contractors completed the two railroad tracks, including

3

the turnout at the south end of the terminal, and progressed the laying of the crane rails. The General Electric Company substituted a No. 6 turnout for the No. 8 called for, in order to make connection to their revised layout of tracks.

All work under the original contract was completed during September, 1918, and the work under the extra work order dated June 4 was completed October 18, 1918.

Terminal Contract No. 103-A

This contract is for furnishing, installing and testing two portable package-freight conveyors for Barge canal terminals. One of these has been installed at Schenectady. The contract was awarded to Brown Portable Conveying Machinery Co., being signed on October 3, 1917. The engineer's preliminary estimate was $3,900.00 each, the contractor's bid, $4,100.00 each. The contract price on this residency as modified by alteration No. 1 is $4,475.80. The value of work done during the year on this residency is $4,100, total done to date, $4,100.00.

F. B. Stoddard, Engineering Assistant, is in charge.

Alteration No. 1, approved by the Canal Board April 16, 1919, provides for furnishing and delivering a portable inclined elevator at Pier 6, New York city, and for changing the carriages of the portable package-freight conveyor. It increases the contract price on this residency by $375.80.

The work done under this contract within the limits of this residency consisted in furnishing, installing and testing a portable package-freight conveyor at the Schenectady terminal. This conveyor arrived October 7, 1918, and was set up and tested by October 12, 1918.

During the first two weeks in June, 1919, the changes called for under alteration No. 1 were made to the conveyor at the Schenectady terminal. To date no estimate has been given on this alteration.

Terminal Contract No. 106

This contract is for furnishing fourteen two-ton steam tractor cranes for Barge canal terminals. It was awarded to the John F. Byers Machine Co., being signed on February 14, 1918. The engineer's preliminary estimate was $5,250.00 per crane, the contractor's bid, $5,265.00 per crane. The contract price as modified by alteration No. 1 is $5,515.00 per crane. The value of work done on this residency during the year is $5,515.00. The work was accepted September 24, 1918, and the final account, amounting on this residency to $11,030.00, was approved by the Canal Board October 9, 1918.

A. P. Mussi, Assistant Engineer, was in charge.

The Schenectady and Amsterdam terminals are included in the list of terminals in this contract. The crane and housing for the Schenectady terminal was delivered during June, 1918. The crane and housing for the Amsterdam terminal was delivered July 9, 1918, but was found to have a broken casting and a sprocket chain missing. These were later replaced and the tractor crane tested on August 2, 1918.

Terminal Contract No. 214 — Amsterdam

This contract is for constructing a frame freight-house and laying pavement at Amsterdam. It was awarded to the Kennedy & Scullen Construction Co., being signed on April 26, 1918. Construction work began May 20, 1918. The engineer's preliminary estimate was $16,478.00, the contractor's bid, $16,323.00. The value of work done during the year is $8,529.00 The work was accepted April 16, 1919, and the final account, amounting to $14,709.00, was approved by the Canal Board April 16, 1919. The amount paid on extra work orders during the year is $335.02, total to date, the same.

A. P. Mussi, Assistant Engineer, was in charge.

An extra work order dated July 22, 1918, provides for removing old and installing new hardware in the easterly freight-house on this terminal. The final account, amounting to $335.02, was approved by the Canal Board March 19, 1919.

The concrete foundation for brick pavement and curbing was progressed during July and August, 1918, and finished early in September. During July the leaching basin was constructed near the east end of the new freight-house and also the carpenter work was practically completed. The roadway and the macadam and brick pavements around the new and old freight-houses were in progress during the summer and completed in October. During October, also, the site was cleaned up.

The electric wiring in the new freight-house was completed and inspected during November, but the contractor was notified to make various changes and this work was not completed until March, 1919.

The work of changing the hardware in the old freight-house under the extra work order dated July 22, 1918, was started the latter part of July and finished during August, 1918.

Terminal Contract No. 12-F — Amsterdam

This contract is for constructing a woven wire fence adjacent to the roadway approach to the Amsterdam terminal. It was awarded to the Anchor Post Iron Works, being signed on April 16, 1918. Construction work began in June, 1918. The engineer's preliminary estimate was $1,289.00, the contractor's bid, $1,379.50. The value of work done during the year is $1,355.25. The work was accepted July 9, 1918, and the final account, amounting to $1,355.25, was approved by the Canal Board July 17, 1918.

A. P. Mussi, Assistant Engineer, was in charge.

Construction work on this contract was completed before June 30, 1918.

Terminal Contract No. 10-P — Fonda

This contract is for paving the terminal site at Fonda. It was awarded to Patrick W. Mulderry, being signed on April 12, 1918. Construction work began June 27, 1918. The engineer's preliminary estimate was $8,602.00, the contractor's bid, $8,700.00. The value of work done during the year is $8,054.40. The work was accepted December 4, 1918, and the final account, amounting to $8,054.40, was approved by the Canal Board January 15, 1919.

A. P. Mussi, Assistant Engineer, was in charge.

The delivery of brick and a small amount of excavation had been done prior to July 1, 1918. The excavation for the brick pavement along the dockwall and for the macadam pavement was done during July and August, 1918. The concrete foundation was placed during August and the brick during September. A small portion of the macadam was placed during September, and in October the macadam north and east of the freight-house was completed and the bottom course placed on the approach to the terminal also a portion of the drainage ditch was excavated. The balance of the work was completed by November 18. Considerable trouble was experienced by the contractor in obtaining and keeping labor on this contract.

Terminal Contract No. 37 — Canajoharie

This contract is for the construction of a harbor and dockwall near the outlet of Canajoharie creek at Canajoharie. It was awarded to Holler & Shepard, being signed on August 26, 1915. Construction work began June 16, 1916. The engineer's preliminary estimate was $33,832.00, the contractor's bid, $32,272.00. The value of work done during the year is $4,936.63. The work was accepted November 19, 1918, and the final account, amounting to $31,436.63, was approved by the Canal Board April 16, 1919. The amount paid on extra work orders to date is $67.71.

T. J. Loonie, Assistant Engineer, was in charge.

During the year concrete pavement has been placed back of the dockwalls, fender timbers have been bolted to the wall and spoil-banks have been graded. Work was completed in October.

Terminal Contract No. 222 — Canajoharie

This contract is for constructing a frame freight-house and depressed roadway at Canajoharie. It was awarded to J. A. Laporte, being signed on August 23, 1918. Construction work began in September, 1918. The engineer's preliminary estimate was $4,000.00, the contractor's bid, $4,195.00. The value of work done during the year is $4,206.40. The work was accepted November 19, 1918, and the final account, amounting to $4,206.40, was approved by the Canal Board February 13, 1919.

H. W. Jewell, Junior Assistant Engineer, was in charge.

The work of building the freight-house and grading the roadway was virtually all done during October, 1918.

Terminal Contract No. 101

This contract is for furnishing and installing steel stiff-leg derricks on terminal sites at Albany, Whitehall, Little Falls, Rome, Lockport and Tonawanda. It was awarded to E. Brown Baker, being signed on December 18, 1916. On February 21, 1917, it was assigned to the Mohawk Dredge & Dock Co., Inc., and this assignment was approved by the Superintendent of Public Works March 26, 1917. The engineer's preliminary estimate was $21,-890.90, the contractor's bid, $31,790.90. Excess metal to the value of $6,510.00 has been authorized by the Canal Board. The work was accepted December 4, 1918, and the final account, amounting to $36,611.92, was approved by the Canal Board December 27, 1918. The amount paid on extra work orders during the year is $2,966.03, total to date, the same, all at Little Falls.

One of these derricks has been installed at Little Falls, the final account for which amounted to $6,011.50.

G. A. Ensign, Assistant Engineer, was in charge.

Final account of extra work order dated December 11, 1917, amounting to $1,776.03 was approved by the Canal Board August 14, 1918.

Final account of extra work order dated April 26, 1918, amounting to $1,190.00 was approved by the Canal Board January 15, 1919.

The derrick at Little Falls was nearly completed at the beginning of the fiscal year. The final estimate was prepared during this year.

Terminal Contract No. 221

This contract is for constructing a frame freight-house at Herkimer and an extension to freight-house at Little Falls. It was awarded to Kennedy & Scullen Construction Co., Inc., being signed on August 30, 1918. The engineer's preliminary estimate was $6,100.00, the contractor's bid, $6,122.00. The value of work done during the year is $5,914.10. The work was accepted March 19, 1919, and the final account, amounting to $5,914.10, was approved by the Canal Board March 19, 1919. The amount paid on extra work orders to date is $379.11.

H. W. Jewell, Junior Assistant Engineer, was in charge.

An extra work order dated October 29, 1918, provides for building partitions in end of freight-house extension at Little Falls. The final account, amounting to $379.11 was approved by the Canal Board January 22, 1919.

This work was all done during the fall of 1918.

Terminal Contract No. 27-P—Frankfort

This contract is for paving the terminal site at Frankfort. It was awarded to Patrick W. Mulderry, being signed on April 12, 1918. The engineer's preliminary estimate was $4,100.00, the contractor's bid, $4,446.00. The value of work done during the year is $3,938.45. The work was accepted November 13, 1918, and the final account, amounting to $3,938.45, was approved by the Canal Board February 19, 1919.

C. W. Wilbur, Junior Assistant Engineer, was in charge.

All of the work was done during the summer and early fall.

Canal Maintenance

REMOVAL OF BARS BELOW LOCK No. 8, AT SCOTIA

On April 13, 1918, the Superintendent of Public Works entered into an agreement with Holler & Shepard for the removal of bars in the Barge canal channel below lock No. 8, at Scotia, payment to be made on a basis of a specified daily rental for plant plus labor and material necessary to carry on the work.

This work was in charge of M. J. Quinn, Junior Assistant Engineer, with office at Schenectady.

In July, 1918, the dipper-dredge belonging to this company continued excavating the gravel bar just below lock No. 8, at Scotia, placing the material along the north bank of the river between Stas. 1345 and 1360. On July 9, the dredge discontinued work at this locality and moved downstream to a point about half way between lock No. 8 and the Scotia trolley bridge, where on the 12th it started removing another bar and continued until August 6, the material being spoiled along the Scotia dike on the north side of the river. After this work was discontinued the dredge left this residency, on August 7, 1918. Subsequent to July 1, 1918, this dredge removed about 8,500 cu. yds. of material from the canal channel.

The cost of this work prior to July 1, 1918, was $8,834.47. During the past year two estimates for this work, amounting to $7,622.61, were approved by this office, making the total cost of the work $16,457.08.

REMOVAL OF BARS BETWEEN FONDA AND INDIAN CASTLE

On April 22, 1918, the Superintendent of Public Works entered into an agreement with the American Pipe and Construction Co., Robert Wetherill, Receiver, for the removal of bars in the Barge canal channel between Fonda and Indian Castle, payment to be made on the basis of a specified daily rental of plant plus labor and material necessary to carry on the work.

The portion of this work located within the limits of this residency was looked after during the past year by Frank S. Belotti, Engineering Assistant, working from the residency office at Amsterdam.

During July, 1918, the excavation of bars was continued by the dipper-dredge at a point about two miles east of lock No. 13, at Yosts, and at a point just west of the Fonda–Fultonville bridge. This material was spoiled along the south bank of the river near Sta. 2840 and along the south bank about a mile east of the Fonda–Fultonville bridge. Considerable time was spent in locating the large boulders it was necessary to remove in the vicinity of the Fonda terminal and in changing the dipper of the dredge for a clam-shell bucket.

On July 29, this dredge moved to the Fonda terminal, where it was dismantled, and on August 1 it was taken to Mindenville to remove a large bar just below lock No. 16· The removal of this bar was completed about the middle of August and then the dredge cleaned the channel downstream from lock No. 16 to a point about one-half mile west of the St. Johnsville bridge. This material was spoiled along the north and south banks of the river about a quarter to a half-mile east of the St. Johnsville bridge. The work in the vicinity of the St. Johnsville bridge was discontinued on September 12, the dredge being dismantled and taken to Indian Castle, to remove a bar in the channel at that place. This bar is located within the limits of what was then residency No. 4.

The work at Indian Castle was finished and the dredge returned to the Mohawk River residency on October 16, being assembled just east of the St. Johnsville bridge, where the work of removing large boulders and high spots in the channel was resumed, the material being spoiled along the banks about one-half mile east of the bridge. This work was suspended on November 2, and the dredge was again dismantled and moved to lock No. 9, to be used there in removing the coffer-dam built for repairs to the dam.

With the removal of the dredge from the vicinity of St. Johnsville, this agreement was terminated. After July 1, 1918, the dredge removed about 14,200 cu. yds. of material from the channel under this agreement.

Prior to July 1, 1918, one estimate for this work, $5,900.44, had been approved by this office. During the past year six additional estimates have been approved, amounting to $31,535.90 and making the total cost of this work $37,436.34.

REMOVAL OF BARS IN THE VICINITY OF CANAJOHARIE

On April 14, 1919, the Superintendent of Public Works entered into an agreement with the American Pipe and Construction Co. for the removal of bars in the Barge canal channel in the vicinity of Canajoharie, payment to be made on the basis of a specified daily rental for plant plus labor and material necessary to carry on the work.

This work has been looked after by Frank S. Belotti, Engineering Assistant, working from the residency office at Amsterdam.

Dipper-dredge No. 3, together with tug and scows left the harbor at Cranesville and arrived at Canajoharie on April 28, 1919. The balance of the month was spent in rigging up the dredge preparatory to removing a large bar in the channel just below dam No. 10 and lock No. 14, at that point. Early in May a cut was made through this bar along the lower guide wall to allow navigation to pass into the lock. The dredge has continued work on this large bar during May and June, removing approximately 21,705 cu. yds. of material to July 1, 1919.

To July 1, 1919, three estimates, totaling $11,600.88, have been approved by this office in connection with this work.

LOCK NO. 10, CRANESVILLE

During the year the Department of Public Works made quite extensive repairs to lock No. 10, at Cranesville. The progress of this work was inspected by forces from this office and reports were sent from time to time to the Division Engineer. Measurements were taken several times at different periods with the lock empty and full, to determine any movement of the river wall of the lock.

A large crack in the floor of the lock along the construction joint parallel with the river wall was discovered. The forces of the Department of Public Works drilled holes in the floor of the lock and in the foundation under the river wall. Voids were found and later were filled with gravel and grout. The crack in the floor of the lock was also filled. The stone in old Erie canal lock No. 26 was removed and hauled to the site of lock No. 10. This stone was then handled by means of a derrick and placed outside of and against the river wall of the lock below the apron of the dam. When the stone work was placed against the wall the joints were all grouted.

No estimates were given on this work as it was done by the forces of the Superintendent of Public Works and was not under the direction of this office except in a general way.

MISCELLANEOUS

During the past year the dam at Scotia has been investigated. The concrete around one of the shoes, against which the upright rests, was found to have been washed away and a large irregular crack was discovered in the concrete apron in the middle span of the dam. The repairs at this dam are being made at the present time by forces of the Superintendent of Public Works.

During the year cross-sections were taken and investigations were made at the other dams on this residency, to determine their condition.

A detailed report of the work necessary for maintenance of the canal during the coming year was made and sent to the Division Engineer.

When navigation opened on this section of the canal for the season of 1919, it was necessary to locate buoys and stake lights for marking the Barge canal channel, and considerable time was devoted to this work by employees of this Department on this residency. During the past year many of the buoys were displaced, owing to floods and other causes. These were replaced by the Department of Public Works with the assistance of the engineers of this Department.

At the close of navigation in 1918 it was estimated that 4,000 cu. yds. of material had been brought down Castle creek and deposited in front of the dam. This material is distributed now along the prism from the guard-gate to the end of the land cut near Rocky Rift dam.

During the winter concrete marker posts, to facilitate finding the offset center line points in the field, were made and the work of placing them has been completed from Mindenville to Little Falls.

Field and Office Work

Estimates of work done and progress of the construction work have been made weekly and monthly to the Division Engineer's office, also estimates prepared for extra work orders and agreements under construction during the year.

Investigations and surveys were made for grants of land under the waters of the Mohawk river at Amsterdam for the Atlas Knitting Company, and also at Schenectady for the American Locomotive Company. Application was received from the American Locomotive Company for several revisions in their maps. These maps were revised and returned to Albany.

The condition of the channel at the Schenectady terminal was investigated and the amount of fill at that point was estimated. The sweeping of the Barge canal channel from lock No. 16, at Mindenville, as far east as lock No. 14, at Canajoharie, was completed during the year. A plan and estimate were made for dredging a channel from the canal prism to the retaining wall along the property of the Chalmer's Knitting Co. at Amsterdam.

An investigation and report were made on the proposed excavation of a slip, or channel, from the Barge canal channel across State land and the old Erie canal to the Cushing Stone Co. property at Cranesville.

An estimate and plan were prepared for a proposed additional widening of the prism below dam No. 5, at Rotterdam.

A survey was made and mapped, showing the blue line intersection with the Amsterdam city lines.

An investigation was conducted and measurements taken to determine the condition of the channel and the maximum clearance between the piers of the New York Central R. R. bridge at Schenectady.

The base line and center line data for Sections Nos. 2 and 3 have been compiled and tabulated on standard forms and sent to the Albany office.

The gages located at Middleburg, Kast Bridge, and the Schenectady Boat Club were reset and changed data sent to the Albany office.

Measurements of controlling dimensions of all locks on this residency and the minimum clearance between the lock gate fen' ers were taken and sent to Albany.

The Sherman complaint, on contract No. 14, and the Eggleston and the Barhydt complaints on contract No. 20-D, were invc' gated as to damages done to these properties, due to flooding.

A survey was made and an appropriation map sent in for the A. Francis Bradt parcel at Rotterdam, on contract No. 185, and a release for spoil was executed covering certain lands of E. J. Millette of Rotterdam, on the same contract.

An appropriation map for lands of Jay Van Evra, on contract No. 20-C, was prepared; also release and retention maps for parcel No. 5171, Lasher, on contract No. 20-C.

An appropriation map of the Sherman property near Vischer Ferry, on contract No. 14, was prepared; also reports and sketches prepared in connection with the release of parcels Nos. 2112, 2023, 2076, 2360 and 2366, on contract No. 14.

An appropriation map of the Watson S. Vrooman lands below Rotterdam was prepared, also a map to supersede parcel No. 3324-A, on contract No. 20-D. Reports and sketches were made in connection with the release of parcels Nos. 3151 and 5132, on contract No. 20-D.

Considerable time has been spent in preparing charts and data to be used in connection with the claim of the American Pipe and Construction Co. on contract No. 20-D.

Release and retention maps were prepared for parcel No. T-50, terminal contract No. 12, at Amsterdam.

The matter of installing a water-supply system for the Guy Park House at Amsterdam, making connection with the city water-main across the N. Y. C. tracks, has been investigated and a report and estimates submitted to the Special Deputy State Engineer.

During the past year some work has been done towards removing bars and obstructions above grade in the Barge canal channel. On July 1, 1919, this work is in progress and will probably be carried on during the coming year.

Within the limits of this residency several claims have been made against the State, due to the eroding of banks. A study of this matter should be made and the banks should be protected in some way in order to prevent future washing away of the banks and consequent damage to lands along the river. This work would necessarily come under maintenance and for the ensuing year the quantities of bank protection would have to be estimated.

At various times during the past year on account of high water and in a few cases for the purpose of making repairs, etc., boatmen have been compelled to tie up along the line of the canal, which lies in river section. In most cases this was done at terminal walls, but where a terminal was not near at hand, it was necessary to use the guide walls of the nearest lock. Although this practice has not as yet congested traffic on the canal, it may do so in the future, especially with increased traffic. I believe, therefore, that a study should be made, looking toward the establishment of suitable mooring places at convenient intervals along the river, so as to provide for increased traffic and for emergencies such as accidents, etc. I take the liberty of mentioning this matter, since I believe it to be a coming necessity in connection with the maintenance of the canal.

Blue Line Surveys

Since fewer men were available than were necessary for the construction work upon the residency, the blue line work was for the most part done during slack times in the construction work. Between October 1, 1918, and April, 1919, there was practically no blue line field work done, except at Little Falls.

At Little Falls the encroachment surveys were made and the encroachments plotted upon the various parcels to be abandoned by the State to the city.

East of Little Falls the following work has been done: The topographic work from Mindenville to Little Falls has been finished, the 40-foot maps as far as Indian Castle have been plotted, and the points staked out in the field and traverses run over them. West of Indian Castle creek the topography has been taken.

The Clinton ditch, or original Erie canal, blue line work has been completed from Mindenville to Indian Castle creek, with the exception of plotting the 100-foot maps.

The Rocky Rift feeder work has been completed and the 100-foot maps sent to Albany.

CHAMPLAIN CANAL, RESIDENCIES NOS. 1, 2 AND 3

Assistant Engineer in charge R. D. Hayes reports:

These residencies cover the entire Champlain canal, Waterford to Whitehall, and include also the terminals on Lake Champlain. The work has been directed from the Mechanicville office.

Numerous release and retention maps have been prepared and reports on various claims submitted to the Division Engineer.

Recording gages, two on the Glens Falls feeder and one at lock No. 9, have been installed.

In connection with Barge canal construction our engineers have been actively engaged in assisting the Department of Public Works, making surveys, sweeping the channel, setting buoys and doing general maintenance work.

Reports follow on contracts Nos. 1-A, 73-A, 131-A and 168 and terminal contracts Nos. 13, 26, 101 and 201.

Contract No. 73-A

This contract is for completing the construction of the canal from Northumberland to Stillwater. Length, 15 miles. It was awarded to the Great Lakes Dredge & Dock Co., being signed on January 15, 1916. The engineer's preliminary estimate was $432,045.00, the contractor's bid, $321,679.92. The contract price as modified by alterations Nos. 1, 2, 3 and 4 is $506,169.67. Excess quantities to the value of $2,270.00 have been authorized by the Canal Board.

C. A. Curtis, Assistant Engineer, and Mott Palmer, Junior Assistant Engineer, were in charge.

Under authority of chapter 585, Laws of 1918, this contract was canceled by the Canal Board on July 9, 1918. The cancelation became effective October 9, 1918, on approval of the Canal Board, the contractor having filed a stipulation of his compliance with the law. The actual cost of the work from April 7, 1917, to

October 9, 1918, was $290,695.71 and payment of balance due on this amount was authorized by the Canal Board on September 17, 1919. The final account for work done prior to April 7, 1917, amounting to $286,278.59, was approved by the Canal Board on May 21, 1919.

This contract has been completed and the total payment, including extra work orders, was $580,344.32.

An extra work order dated September 5, 1916, provided for placing six timber snubbing-posts above guard-lock No. 10. The final account, amounting to $247.55, was approved by the Canal Board May 21, 1919.

The contractor, using equipments of the American Pipe and Construction Co., the Troy Public Works Co. and Hawley Miller in addition to his own plant, completed prism excavation between locks Nos. 4 and 5 by October 9, 1918.

The snubbing-posts called for by the work order of September 5, 1916, were placed.

Contract No. 168

This contract is for constructing concrete-capped timber guide-cribs near locks Nos. 3, 5 and 6, Champlain canal. It was awarded to Bronk and Kimmey, being signed on July 29, 1918. The engineer's preliminary estimate was $63,505.60, the contractor's bid, $77,895.12. The contract price as modified by alteration No. 1 is $85,727.40. The value of work done during the year is $67,240, total done to date, the same. The amount paid on extra work orders during the year is $2,555.48, total to date, the same.

C. A. Curtis, Assistant Engineer, is in charge.

Alteration No. 1, approved by the Canal Board March 19, 1919, provides for three additional cribs below lock No. 3, Mechanicville. It increases the contract price by $7,832.28.

An extra work order dated September 24, 1918, provides for restoring cribs displaced by boats. The final account, amounting to $2,555.48, was approved by the Canal Board January 22, 1919.

An extra work order dated May 29, 1919, provides for placing rubbing strips on piers below lock No. 3, Mechanicville.

The 15 cribs called for below lock No. 3, were completed during 1918. The timber cribs with stone filling above lock No. 5 were also placed in 1918. Timber for the cribs below lock No. 6 was provided. Under alteration No. 1 three additional cribs were placed below lock No. 3. Concrete work and the building of booms was started above lock No. 5 and below lock No. 6.

Previous to the placing of the concrete caps on the cribs below lock No. 3, an ore fleet from Lake Champlain struck two of these cribs and knocked them off their location. To place them back in line the contractors were given an extra work order, under which the work was done.

Contract No. 131-A

This contract is for completing the reconstruction of the highway bridge crossing the main channel of the Hudson river at Schuylerville. It was awarded to Michael Fitzgerald, being signed on March 5, 1917. Construction work began in April, 1917. The engineer's preliminary estimate was $30,753.00, the contractor's bid, $39,634.50. Excess foundation piles to the value of $140.00 have been authorized by the Canal Board. The work was accepted July 17, 1918, and the final account, amounting to $37,276.28, was approved by the Canal Board September 10, 1918. The amount paid on extra work orders to date is $640.04.

C. A. Curtis, Assistant Engineer, was in charge.

Construction work was completed prior to July 1, 1918. The contract was accepted and the final account approved during the past fiscal year.

Contract No. 1-A

This contract is for completing the canal from Crocker's Reef to Fort Edward. It was awarded to Holler & Shepard, being signed on August 31, 1914. The engineer's preliminary estimate was $90,811.00, the contractor's bid, $120,459.40. The contract price as modified by alterations Nos. 1, 2 and 3 is $141,540.20. Excess excavation to the value of $67,200.00 has been authorized by the Canal Board.

James B. Foote, Assistant Engineer, was in charge.

Under authority of chapter 585, Laws of 1918, this contract was canceled by the Canal Board on July 24, 1918. The cancelation became effective August 14, 1918, on approval of the Canal Board, the contractor having filed a stipulation of his compliance with the law. The actual cost of the work from April 7,

1917, to August 14,1918, was $44,663.42 and payment of balance
due on this amount was authorized by the Canal Board on April
2, 1919. The final account for work done prior to April 7, 1917,
amounting to $166,941.19, was approved by the Canal Board on
April 2, 1919.

This contract has been completed and the total payment, including extra work orders, was $212,254.61.

An extra work order dated March 29, 1917, provides for fur-
nishing, placing and removing flash-board at Crocker's Reef dam.
The final account, amounting to $650.00, was approved by the
Canal Board August 14, 1918.

The only contract work done during the year was that performed under the work order of March 29, 1917.

Terminal Contract No. 13 — Schuylerville

This contract is for constructing a guard-lock and bridge at
Schuylerville. It was awarded to Lou B. Cleveland, being signed
on December 29, 1914. It was assigned to the Kendar Engineering
and Construction Co., Inc., and this assignment was approved by
the Superintendent of Public Works December 16, 1915. The
engineer's preliminary estimate was $61,664.60, the contractor's
bid, $42,742.80.

The status of this contract has not changed during the past
year. Construction work has been finished, the final estimate has
been made, but the settlement is still in litigation.

Terminal Contract No. 201 — Whitehall

This contract is for constructing terminal warhouses at Albany
and Whitehall. It was awarded to J. A. Laporte, being signed
on January 2, 1917. The engineer's preliminary estimate was
$59,300.00, the contractor's bid, $65,174.85. The contract price
as modified by alteration No. 1 is $71,432.30. The engineer's
preliminary estimate for the warehouse at Whitehall was $22,-
800.00, the contractor's bid, $28,068.35.

W. L. Caler, Assistant Engineer, is in charge.

Under authority of chapter 585, Laws of 1918, this contract
was canceled by the Canal Board on July 17, 1918. The cancelation became effective August 31, 1918, on approval of the

Canal Board, the contractor having filed a stipulation of his compliance with the terms of the law. The actual cost of the work from April 7, 1917, to August 31, 1918, was $42,790.16 and payment of balance due on this amount was authorized by the Canal Board on February 26, 1919. No work was done prior to April 7, 1917.

The contractor is completing the work and the total payments to date, including extra work orders, are $54,234.00.

At the Whitehall warehouse during the year excavation for the depressed roadway was made and fenced. The side walls were completed and stuccoed. The tile was placed and the chimney was completed. The partitions were put in, the walls were plastered, the metal ceiling was built over the second floor and the stairs and railing around the stair well were placed. The concrete floors were laid and the walls painted in office and warehouse. Metal trim was placed in the office. The skylights, the entrance doors, and the windows and doors in both warehouse and office were placed, and the copper roof was put on the marquise. Plumbing and electrical wiring were completed and a hot-water heating system installed. A 4-inch pipe line was laid to connect the building with the city main. Fire hose, brackets and valves were installed on the fire risers.

Terminal Contract No. 101

This contract is for furnishing and installing steel stiff-leg derricks on terminal sites at Albany, Whitehall, Little Falls, Rome, Lockport and Tonawanda. It was awarded to E. Brown Baker, being signed on December 18, 1916. On February 21, 1917, it was assigned to the Mohawk Dredge and Dock Co., Inc., and this assignment was approved by the Superintendent of Public Works March 26, 1917. The engineer's preliminary estimate was $21,890.90, the contractor's bid, $31,790.90. Excess metal to the value of $6,510.00 has been authorized by the Canal Board. The work was accepted December 4, 1918, and the final account, amounting to $36,611.92 was approved by the Canal Board December 27, 1918. The amount paid on extra work orders during the year was $2,966.03, total to date, the same.

One of these derricks has been installed at Whitehall, the final account for which amounted to $6,381.16.

W. L. Caler, Assistant Engineer, was in charge.

Terminal Contract No. 26 — Rouses Point

This contract is for constructing a pier and basin at Rouses Point. It was awarded to John E. Byron & Co., being signed on October 30, 1916. The engineer's preliminary estimate was $51,200.00, the contractor's bid, $55,678.50.

Under authority of chapter 585, Laws of 1918, this contract was canceled by the Canal Board on July 9, 1918. The cancelation became effective October 9, 1918, on approval of the Canal Board, the contractor having filed a stipulation of his compliance with the terms of the law. The actual cost of the work from April 7, 1917, to October 9, 1918, was $23,456.67 and payment of balance due on this amount was authorized by the Canal Board on March 19, 1919. No work was done prior to April 7, 1917.

Operations were suspended on December 21, 1917, and no work has been done since that time.

Canal Maintenance

Under a special agreement between D. B. La Du and the Superintendent of Public Works, dated April 16, 1919, for dredging the canal and removing high spots south of Fort Edward, 17,800 cubic yards of material were removed.

Under a special agreement between Dwight B. La Du Company and the Superintendent of Public Works, dated May 14, 1919, and June 3, 1919, for placing riprap at Collins island, below Fort Edward, and at other places, the work at Fort Edward was completed and that at Collins island begun.

Under an agreement between J. W. Holler and the Superintendent of Public Works, dated May 29, 1918, for dredging the channel of the Barge canal between lock No. 8 and the Comstock bridge, 27,705 cubic yards of material were removed.

Under agreements between J. W. Holler and the Superintendent of Public Works, dated April 3, 1919, and May 22, 1919, for dredging bars from lock No. 8 to Whitehall and cleaning locks, 4,620 cubic yards of material were removed.

Under an agreement between D. B. La Du and the Superintendent of Public Works, dated May 9, 1919, for removing a ledge of rock in the vicinity of the foot-bridge at Whitehall, rock drilled and blasted.

New York Residency

Senior Assistant Engineer Edward Amderberg reports:

The principal work of the New York residency for the past year has been the supervision of construction work relating to the improvement of Barge canal terminals in this city. Progress was made towards the completion of two terminals for the borough of Manhattan, two for Brooklyn, one for the Bronx and one for Queens. Plans were prepared for developing the two additional sites owned by the State in the borough of Queens.

In addition to the work connected with the Barge canal terminals, various miscellaneous engineering matters were looked after, as decided in the following five paragraphs:

The New York–New Jersey Port and Harbor Development Commission in its studies for improving port conditions in New York harbor decided to investigate the feasibility and cost of constructing a harbor belt line railroad which would intersect the various trunk line railroads in New Jersey some distance inland from the existing terminals. At the request of this Commission, this survey was undertaken by the State Engineer and Surveyor. The survey was started in September, 1918, and was continued to March, 1919, when it was necessary to terminate it temporarily because of lack of funds in the hands of the Commission. The field survey has been completed sufficiently to locate a center line from Piermont on the Hudson to Newark bay, a distance of 45 miles. It is is expected that the remaining work, which amounts principally to the completion of the maps, will be undertaken the coming year. J. O. Burt and E. H. Anderson, Assistant Engineers, acted as locating engineers on this work.

In connection with the investigations of the Jamaica Bay–Peconic Bay Canal Board, under chapter 317, Laws of 1917, for a proposed canal along the south shore of Long Island, a survey was made for a third alternative line across the Rockaway peninsula known as the " Far Rockaway route." A survey for a fourth alternative line, known as the " Lynbrook route," is under way.

Under chapter 427, Laws 1918, a survey of Mill river in Nassau county was made.

Surveys have been made and maps prepared in connection with five applications for grants of land under water in this vicinity.

Plans were prepared for repairs to a pier at the Swinburne Island State quarantine station.

Reports on terminal contracts Nos. 19, 38, 42, 44, 44-P, 52, 52-P, 55, 56, 77, 102, 103-A, 105, 117, 207, 207-P, 207-H, 217, 217-P, 218, 223 and 227 and parts of terminal contracts Nos. 106 and 109 follow; also on contracts for a highway at Rockaway Point, repairs to the landing pier and the north pier at the Hoffman Island State quarantine station and repairs to sea-walls between East Marion and Orient.

Terminal Contract No. 52 — Pier 6, East River

This contract is for repairing and extending existing Pier 6, East river. It was awarded to Kaufman & Garcey, being signed on July 27, 1916. Construction work began October 13, 1916. The engineer's preliminary estimate was $89,974, the contractor's bid, $91,317.75. The contract price as modified by alterations Nos. 1, 2, 3 and 4 is $102,553.75. The amount paid on extra work orders during the year is $130, total to date, $2,856.04.

Ely Gamse, Assistant Engineer, was in charge.

Under authority of chapter 585, Laws of 1918, this contract was canceled by the Canal Board on July 9, 1918. The cancelation became effective August 31, 1918, on approval of the Canal Board, the contractor having filed a stipulation of his compliance with the terms of the law. The actual cost of the work from April 7, 1917, to August 31, 1918, was $100,656.77 and payment of balance due on this amount was authorized by the Canal Board on February 13, 1919. The final account for work done prior to April 7, 1917, amounting to $31,491.89, was authorized by the

Canal Board on January 29, 1919. The total payments, including extra work orders, on the completed contract were $135,004.70.

An extra work order dated December 7, 1917, provided for replacing stone block payment back of the bulkhead wall. The final account, amounting to $130.00, was approved by the Canal Board September 24, 1918.

Work on the contract was started October 13, 1916, and had been practically completed at the close of the last fiscal year. During July and August, 1918, some remaining bolts in low-water timber connections were placed at favorable tides and other miscellaneous work was done. On August 28, 1918, all work had been completed.

Terminal Contract No. 103-A

This contract is for furnishing, installing and testing two portable package-freight conveyors. It was awarded to the Brown Portable Conveying Machinery Co., being signed on October 3, 1917. The engineer's preliminary estimate was $3,900.00 each, the contractor's bid, $4,100.00 each. The contract price, on this residency, as modified by alteration No. 1 is $5,626.80.

Ely Gamse, Assistant Engineer, is in charge.

Alteration No. 1, approved by the Canal Board April 16, 1919, provides for furnishing and delivering a portable inclined elevator at Pier 6 and for changing the carriages of the portable package-freight conveyors. It increases the contract price, on this residency, by $1,526.80.

One of the portable package conveyors and the portable package inclined elevator were delivered at Pier 6 during June, 1919. These machines require adjustment and repairs before acceptance tests can be made.

Terminal Contract No. 106

This contract is for furnishing fourteen two-ton steam tractor cranes for Barge canal terminals. Two of these are for Pier 6 and one for Long Island City. The contract was awarded to John F. Byers Machine Co., being signed on February 14, 1918. The engineer's preliminary estimate was $5,250.00 per crane, the contractors bid, $5,265.00 per crane. The contract price as modified by alteration No. 1, is $5,515.00 per crane. The value of the work done at Pier 6, East river, New York city, during the year is $1,370.00. The contract was accepted September 24, 1918, and the final account, amounting to $11,030.00 for Pier 6, was approved by the Canal Board October 9, 1918.

Ely Gamse, Assistant Engineer, was in charge.

Since the last report two clam-shell buckets in connection with the two cranes for Pier 6 were delivered. The crane for the Long Island City terminal was delivered temporarily at Little Falls, for use there, pending a demand for it in this city. The transfer of the latter crane to New York was never ordered, but the work was accepted in September, 1918, and the contract closed.

Terminal Contract No. 207 — Pier 6, East River

This contract is for constructing a freight-shed and head-house on Pier 6, East river. It was awarded to I. J. Stander & Co., Inc., being signed on January 11, 1918. Construction work began in December, 1918. The engineer's preliminary estimate was $133,500.00, the contractor's bid, $128,250.01. The value of work done during the year is $113,390.00, total done to date, the same.

Ely Gamse, Assistant Engineer, is in charge.

Work was started in the field the last of December, 1918. To date the contractor has erected practically all the steelwork. The remaining work comprises the completion of doors and windows and considerable work in connection with the partitions for the offices of the head-house.

Terminal Contract No. 207-P — Pier 6, East River

This contract is for installing a plumbing and water-supply system in the freight-shed on Pier 6, East river. It was awarded to Jarcho Bros., Inc., being signed on April 16, 1918. The engineer's preliminary estimate was $6,000.00, the contractor's bid, $6,650.00. The value of work done during the year is $5,190.00, total done to date, the same.

Ely Gamse, Assistant Engineer, is in charge.

Work was started on this contract in the middle of March, 1919. All piping for the plumbing and water-supply systems has been completed. The remaining work involves the placing of the fixtures, which will not be installed until the completion of plastering, etc., by the contractor on terminal contract No. 207.

Terminal Contract No. 207-H — Pier 6, East River

This contract is for installing a heating system in the freight-shed on Pier 6, East river. It was awarded to Miller & Brady, Inc., being signed on March 22, 1918. The engineer's preliminary estimate was $3,250.00, the contractor's bid, $2,352.00. The value of work done during the year is $1,170.00, total done to date, the same.

Ely Gamse, Assistant Engineer, is in charge.

Work on this contract was started on June 12, 1919. To date the boiler has been installed, some of the piping has been assembled and some of the hangers for ceiling radiators placed.

Terminal Contract No. 52-P — Pier 6, East River

This contract is for paving Pier 6, East River, New York city. It was awarded to the Sicilian Asphalt Paving Co., being signed on June 13, 1919. The engineer's preliminary estimate was $10,000.00, the contractor's bid, $11,225.00.

Work has not yet begun.

Terminal Contract No. 105 — Pier 6, East River

This contract is for installing electric equipment for light, power and battery-charging, and auto-truck scales at Pier 6, East river. It was awarded to the Lord Electric Co., being signed on June 28, 1918. The engineer's preliminary estimate was $17,-742.40, the contractor's bid, $16,000.50. The value of work done during the year is $8,370.00, total done to date, the same.

Ely Gamse, Assistant Engineer, is in charge.

An extra work order dated July 15, 1918, provides for running a feeder line for electric service on Pier 5, East river. The final account, amounting to $2,842.68, was approved by the Canal Board May 7, 1919.

An extra work order dated April 26, 1919, provides for raising the metal ducts along the east side of the pier-shed so as to avoid interference with the opening of the warehouse doors. The final account, amounting to $98.75, was approved by the Canal Board May 21, 1919.

Work on this contract was started on July 9, 1918, in connection with the extra work order providing for running a feeder line out on Pier 5 from Pier 6. This extra work was completed in December. Work on the original contract was started early in February of this year. To date practically all of the rigid metal conduit and outlet boxes have been placed. A considerable proportion of the conductors have been pulled through the conduits. Some of the lighting and power fixtures have been placed. The auto-truck scales have been installed. The switchboard and the balancer sets have not been installed as yet.

Terminal Contract No. 109

This contract is for furnishing electric capstans and trolley hoists at Pier 6, East river, and West 53d street pier, New York city, and electric capstans at the Utica terminal lock. It was awarded to the General Electric Co., being signed on August 2, 1918. The engineer's preliminary estimate was $19,000.00, the contractor's bid, $18,536.00. The value of work done during the year is $1,170.00, total done to date, the same. The amount paid on extra work orders during the year is $219.00, total to date, the same.

Ely Gamse, Assistant Engineer, is in charge.

An extra work order dated September 24, 1918, provides for furnishing a motor with starter, pulley and base for a portable conveyor at Pier 6. The final account, amounting to $219.00, was approved by the Canal Board November 13, 1918.

One winch has been delivered at Pier 6.

Terminal Contract No. 102 — Pier 6, East River

This contract is for furnishing and installing two 3-ton, electric, semiportal, revolving, jib cranes on Pier 6, East river. It was awarded to the Edward F. Terry Mfg., Co., being signed on January 21, 1919. The engineer's preliminary estimate was $45,362.00, the contractor's bid, $41,081.00.

Shop work is under way.

Terminal Contract No. 56 — Pier 5, East River

This contract is for repairing Pier 5, East river. It was awarded to I. J. Stander & Co., Inc., being signed on June 28, 1918. Construction work began June 28, 1918. The engineer's preliminary estimate was $20,400.00, the contractor's bid, $27,-159.60. Excess iron castings to the value of $10.40 have been authorized by the Canal Board. The value of work done during the year is $26,904.68, total done to date, the same. The work was accepted February 26, 1919, and the final account, amounting to $26,904.68, was approved by the Canal Board June 11, 1919. The amount paid on extra work orders during the year is $2,889.71, total to date, the same.

Ely Gamse, Assistant Engineer, is in charge.

An extra work order dated October 5, 1918, provides for jacking up range timbers, placing hardwood shims under same and also for replacing existing posts which are not of sufficient length. The final account, amounting to $2,889.71, was approved by the Canal Board May 7, 1919.

The repair work involved the repair of the ends of the timber bents an the sides of the pier where they had decayed and the placing of an entire new fender system. The required work was completed the latter part of February, 1919.

Terminal Contract No. 19 — Greenpoint

This contract provides for dredging the terminal basin, constructing a new reinforced concrete bulkhead wall and pier, and repairing two existing timber piers and a timber bulkhead at the Greenpoint terminal, borough of Brooklyn. It was awarded to McHarg-Barton Co., being signed on November 24, 1916. Construction work began February 20, 1916. The engineer's preliminary estimate was $193,500.00, the contractor's bid, $207,-383.00. The contract price as modified by alterations Nos. 1 and 2 is $211,513.00. Excess quantities to the value of $1,835.00 have been authorized by the Canal Board. The amount paid on extra work orders to date is $9,356.80.

J. B. Doughty, Assistant Engineer, was in charge.

Under authority of chapter 585, Laws of 1918, this contract was canceled by the Canal Board on July 9, 1918. The cancelation became effective August 14, 1918, on approval of the Canal Board, the contractor having filed a stipulation of his compliance with the terms of the law. The actual cost of the work from April 7, 1917, to August 14, 1918, was $218,767.51 and payment of balance due on this amount was authorized by the Canal Board on January 15, 1919. The final account for work done prior to April 7, 1917, amounting to $6,312.78, was authorized by the Canal Board on December 27, 1918.

The total payments on the completed work, including extra work order and the work done by the State Engineer, amount to $292,191.18.

At the close of the last fiscal year the dredging, the construction of the reinforced concrete bulkhead and the repairs to the Dupont street pier had been practically all completed. The reinforced concrete pier had been 40 per cent completed and the repairs to the old timber crib were well under way.

A considerable amount of work remained to be done after August 14, 1918, and most of it was performed through the contractor on a basis of the actual and necessary cost and expense as prescribed in chapter 585, Laws of 1918. By December 1, 1918, only a small amount of work remained to be done, but owing to a strike of the dock-builders in this city, the work through the contractor was discontinued. The payments to the contractor for work done after August 14, 1918, aggregated $55,225.96.

The small amount of work remaining to be done on December 1 was performed during January and February by the State Engineer under resolution of the Canal Board dated December 27, 1918, authorizing him to purchase the necessary material and hire the necessary plant and labor at a cost not to exceed $3,000. By February 28, 1919, the work under the State Engineer, amounting to $2,528.13, had been completed.

Terminal Contract No. 223 — Greenpoint

This contract is for constructing a terminal freight-shed at Greenpoint, New York city. It was awarded to Post & McCord, being signed on March 20, 1919. The engineer's preliminary estimate was $99,710.00, the contractor's bid, $75,718.94.

J. B. Doughty, Assistant Engineer, is in charge.

An extra work order dated April 1, 1919, provides for repairing an existing concrete warehouse on the site.

Work in connection with the repair of the existing warehouse under the extra work order was started on March 28, 1919, and was completed by the first of June.

Work in connection with the original contract has not been started.

Terminal Contract No. 55 — Gowanus Bay

This contract is for constructing a pier, 150 ft. by 1,220 ft., dolphins, etc., at Gowanus bay terminal. It was awarded to River-side Contracting Co., being signed on September 4, 1917. Construction work began October 5, 1917. The engineer's preliminary estimate was $513,000.00, the contractor's bid, $509,800.75. The contract price as modified by alterations Nos. 1, 2 and 3 is $507,620.40. The value of work done during the year is $102,750, total done to date, $260,080.

L. T. Howard, Assistant Engineer, is in charge.

Alteration No. 2, approved by the Canal Board January 15, 1919, provides for eliminating the scale pits, for changing the type of capstan settings, for changing the details of the anchorage for the horizontal clamps for the shed footings on the west side of the pier, and for installing conduits and manholes for electric wires. It decreases the contract price by $780.35.

Alteration No. 3, approved by the Canal Board June 25, 1919, provides for eliminating from the site of the contract a portion of the terminal area back of the bulkhead wall. It does not change the contract price.

An extra work order dated May 23, 1919, provides for moving part of the stored materials on the contractor's site to allow stone block pavement to be laid under terminal contract No. 44-P.

The work to date on this contract includes the completion of driving and framing all foundation piles except considerable work remaining to be done in framing the piles under the freight-house and crane rail foundations. Bracing piles have been driven, but considerable framing remains to be done on them. About one-third of the fender piles have been driven. The timber work has been largely completed. The principal lumber still to be placed is in the fender system, the sheathing at the outer end, the grillage timbers and chocks in the foundations for the concrete blocks, the fender pile caps, etc., of the pier, the pile dolphins and the east pier seat.

The contractor suffered considerable delay because lighters and barges engaged in U. S. Government work trespassed on his site. His work was also held up from November 20, 1918, to January 6, 1919, by a strike of the dock-builders in New York city.

The contractor's plant consists of a floating pile-driver derrick lighter, compressor boat, tanks for timber treatment, etc.

Terminal Contract No. 227 — Gowanus Bay

This contract is for constructing a frame freight-house at Gowanus bay. It was awarded to J. A. Laporte, being signed on April 28, 1919. The engineer's preliminary estimate was $9,750.00, the contractor's bid, $9,189.00. Excess second-class concrete to the value of $300.00 has been authorized by the Canal Board. The value of work done during the year is $8,000.00, total done to date, the same.

L. T. Howard, Assistant Engineer, is in charge.

Work was started on April 23, 1919, and by the end of the fiscal year the carpenter work had been practically completed, a priming coat of paint put on and the electric lighting system partly installed.

Terminal Contract No. 44 — Mott Haven

This contract is for excavating a terminal basin, constructing a dockwall, grading the upland and building approaches for a Barge canal terminal at Mott Haven, near East 138th street, borough of Bronx. It was awarded to Geo. W. Rogers & Co., Inc., being signed on June 8, 1917. Construction work began June 9, 1917. The engineer's preliminary estimate was $170,300.00, the contractor's bid, $193,651.00. The contract price as modified by alteration No. 1 is $191,195.50. The value of work done during the year is $55,370.73. The work was accepted Feburay 26, 1919, and the final account, amounting to $176,110.73, was approved by the Canal Board August 6, 1919.

F. T. Lawton, Assistant Engineer, was in charge.

An extra work order dated February 24, 1919, provides for laying a 6-inch water-main from the street line to the site of the proposed freight-house.

An extra work order, dated March 8, 1919, provides for remodeling an existing building for warehouse purposes.

During the past year the work has comprised the completion of the bulkhead wall, the completion of the upland excavation, the construction of the street and ramp walls, the repairs to the existing crib at the slip and the construction of the steel sheet pile wall. This work was completed by the latter part of February of this year. The repairs to the existing brick building were started the first of March, 1919, and carried to completion by the middle of June.

The contractor's plant consisted of two floating pile-drivers, a derrick-scow, a compressor, etc.

Terminal Contract No. 38 — West 53d Street, North River

This contract is for constructing a pier at the foot of West 53d street, North river. It was awarded to I. J. Stander & Co., Inc., being signed on October 27, 1917. Construction work began November 1, 1917. The engineer's preliminary estimate was $259,000.00, the contractor's bid, $266,064.80. The contract price as modified by alterations Nos. 1 and 2 is $265,550.39. The value of work done during the year is $162.900, total done to date, $214,390. The amount paid on extra work orders during the year is $65.00, total to date, $470.00.

W. C. Bratton, Assistant Engineer, is in charge.

Alteration No. 2, approved by the Canal Board June 25, 1919, provides for the elimination of a part of the backing log and for revised settings for single bitts. It increases the contract price by $47.23.

An extra work order dated August 1, 1918, provides for painting the structural steel delivered and not used and for transporting and placing it in storage at the Greenpoint terminal warehouse. The final account, amounting to $65.00, was approved by the Canal Board September 10, 1918.

An extra work order dated September 13, 1918, provides for altering the crane rail.

An extra work order dated February 5, 1919, provides for additional column footings on the bulkhead wall for the proposed head-house.

During the past year the work has included the driving of all remaining foundation and bracing piles. Fender piles have been 80 per cent driven and framed. Timber framing was very largely completed. The remaining timber work is principally the placing of caps and grillage timbers for concrete footings and can be placed only at low tides, making rapid progress difficult. The concrete footings have been partly built and the crane rail girder partly erected. The concrete deck covering is still to be placed.

The contractor's plant consisted of a floating pile-driver, a compressor plant, two concrete mixers and tanks for treatment of lumber.

Terminal Contract No. 218 — West 53d Street

This contract is for constructing a terminal freight-shed at West 53d street, New York city. It was awarded to Donnell-Zane Co., Inc., being signed on May 13, 1919. The engineer's preliminary estimate was $53,969.15, the contractor's bid, $46,549.20.

Construction work has not yet begun.

Terminal Contract No. 217 — Long Island City

This contract is for repairing the bulkhead and constructing a freight-house and a crane track at Long Island City. It was awarded to A. E. Norton, Inc., being signed on October 23, 1918. Construction work began December 2, 1918. The engineer's preliminary estimate was $59,850.00, the contractor's bid, $74,806.50. Excess quantities to the value of $5.780.00 have been authorized by the Canal Board. The value of work done during the year is $48,860.00, total done to date, the same. The amount paid on extra work to date is $150.00.

H. W. Hale, Assistant Engineer, is in charge.

An extra work order dated December 27, 1918, provides for furnishing and erecting a 10 ft. by 16 ft. wooden building. The final account, amounting to $150.00, was approved by the Canal Board February 26, 1919.

An extra work order dated May 7, 1919, provides for furnishing and installing a heating plant in the freight-house.

The work done on this contract includes a small percentage of the bulkhead repairs, the completion of the foundation of the freight-house and the erection of the steelwork, which has been 75 per cent riveted.

Terminal Contract No. 217-P — Long Island City

This contract is for installing plumbing and water-supply systems in the terminal freight-house at Long Island City. It was awarded to the Altman Plumbing Co., being signed on May 29, 1919. The engineer's preliminary estimate was $4,000.00, the contractor's bid, $3,765.00. The value of work done during the year is $1,390.00, total done to date, the same.

H. W. Hale, Assistant Engineer, is in charge.

The plumbing installation was started on June 13 and to date the trenching for the water-pipe supply line and for the sewer have been completed and the 4-inch and 6-inch water-pipe lines have been partly installed.

Terminal Contract No. 42 — Long Island City

This contract is for paving the terminal site at Long Island City. It was awarded to Leonard Paving Co., Inc., being signed on November 13, 1918. The engineer's preliminary estimate was $54,600, the contractor's bid, $53,579.00. The value of work done during the year is $1,370.

H. W. Hale, Assistant Engineer, is in charge.

Grading was started on March 15, 1919, and to date a certain amount of it has been completed around the freight-house. Also some tile drain has been laid. The work was discontinued temporarily on April 9, pending the completion of the repairs to the bulkhead and the construction of the crane track under terminal contract No. 217.

Terminal Contract No. 44-P

This contract is for paving parts of terminal sites at Mott Haven, Greenpoint and Gowanus bay. It was awarded to the Asphalt Construction Co., being signed on May 13, 1919. The engineer's preliminary estimate was $94,340.00, the contractor's bid, $81,360.00. The contract price as modified by alteration No. 1 is $78,201.20. The value of work done during the year is $22,070, total done to date, the same.

FREIGHT-SHED, GREENPOINT TERMINAL, NEW YORK CITY

In addition to this shed there was an existing building at this site which has been altered so as to adapt it for use for storage purposes.

Interior of Freight-shed, Greenpoint Terminal, New York City

FRAME FREIGHT-HOUSE AT GOWANU'S BAY TERMINAL

The permanent freight-house is to be erected on the pier and will be 1,180 feet long by 106 feet wide and have cargo masts and a conveyor gallery along one side.

MU

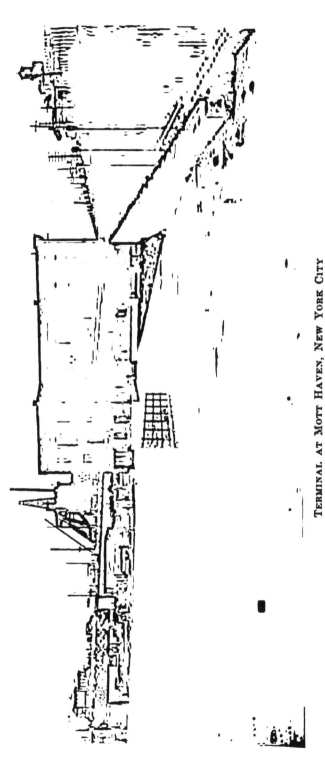

TERMINAL AT MOTT HAVEN, NEW YORK CITY

View looking down ramp leading to terminal area. The existing brick building, which has been remodeled so as to adapt it for use as a storage warehouse, is seen in the middle distance.

MU

WEST 53D STREET TERMINAL, NEW YORK CITY

Erection of pier-shed. Construction of the pier has been in progress during the year.

F. T. Lawton, Assistant Engineer, is in charge at Mott Haven, L. T. Howard, Assistant Engineer, is in charge at Gowanus and J. B. Doughty, Assistant Engineer, is in charge at Greenpoint.

Alteration No. 1, approved by the Canal Board June 25, 1919, provides for elimination of the crane track and the pavement between the freight-house site and the dockwall and for paving the entire area west of the approach driveway at Mott Haven. It decreases the contract price by $3,158.80.

Work was started at the Mott Haven site on May 28, 1919, at the Gowanus site on June 2, and at the Greenpoint site on June 30. At Mott Haven the grading has been completed and 89 per cent of the concrete pavement base has been laid, including concrete edging. At Gowanus the excavation has been completed and practically all the concrete base and edging have been laid.

Terminal Contract No. 77

This contract is for dredging at Piers 5 and 6, East river, at Greenpoint and at Long Island City. It was awarded to New Jersey Shipbuilding & Dredging Co., being signed on May 13, 1919. Construction work began June 13, 1919. The engineer's preliminary estimate was $40,325.00, the contractor's bid, $42,895.00. The value of work done during the year is $6,770.00, total done to date, the same.

H. W. Hale, Assistant Engineer, is in charge at Long Island City.

A dredge started work at the Long Island City terminal on June 13, 1919, on the outside cut in front of the bulkhead wall. By July 1, a total of 4,926 cubic yards of material had been removed and dumped at sea. The dredging has not yet been started at the other sites.

Terminal Contract No. 117

This contract is for furnishing battery-charging motor-generator sets, with switchboard panels for New York city terminals. It was awarded to The Electric Products Co., being signed on June 20, 1919. The engineer's preliminary estimate was $7,800.00, the contractor's bid, $5,292.52.

No work has yet been done.

4

Highway at Rockaway Point, L. I.

Under chapter 130, Laws of 1917, a concrete highway was constructed adjacent to Fort Tilden, Rockaway Point, L. I.

The contract was awarded to the Rosoff Engineering Co., Inc., of New York city, for $44,126.25, which was subsequently increased to $46,126.25 under a supplementary agreement providing for extending the pavement across the property of the U. S. Coast Guard. The work on this road was started the first of July, 1918, and completed in October, 1918. The final account was $45,989.95.

H. W. Hale, Assistant Engineer, was in charge.

Repairs to Landing and North Piers, Hoffman Island, State Quarantine Station

In connection with repairs to the landing pier and to the north pier, Hoffman Island, State Quarantine Station, two contracts were supervised by this Department, for the Health Officer of the Port of New York. The work on the landing pier was done during the previous year, but the final account was computed during the current year.

The landing pier was repaired under a contract let January 3, 1918, to A. M. Hazell, Inc., of New York city, for $3,387.00. Under an extra work order dated May 21, 1918, the extent of the repairs was increased, for which payment was to be made on a basis of cost plus 15 per cent. The work was started early in April, 1918, and completed on June 19, 1919. The final account for the original contract was $4,152.00 and for the extra work, $3,047.61.

The north pier was repaired under a contract let June 26, 1918, to Anderson and Wheeler, Inc., of New York city, on a basis of cost plus 10 per cent. The extent of the work contemplated was subsequently extended and the additional work was done under an extra work order dated August 9, 1918, on a basis of cost plus 10 per cent. The final account of the original contract was $4,562.02 and for the extra work, $4,354.44.

L. T. Howard, Assistant Engineer, was in charge.

Repairs to Sea-walls between East Marion and Orient, L. I.

Under chapter 428, Laws of 1918, general repairs were made to two existing sea-walls for retaining a highway connecting the villages of East Marion and Orient, town of Southold, Suffolk county. The contract was let to the Rosoff Engineering Co., Inc., of New York city on October 14, 1918, at a bid price of $8,858.50. Work was started on October 24 and completed on December 4, 1918. The final account was $8,818.81.

H. W. Hale, Assistant Engineer, was in charge.

THE FOLLOWING STATEMENTS SHOW THE NAMES, RANK AND COM-
PENSATION OF ENGINEERS EMPLOYED IN THE EASTERN DIVISION
OF THE DEPARTMENT OF THE STATE ENGINEER AND SURVEYOR,
TOGETHER WITH INCIDENTAL EXPENSES, FOR THE FISCAL YEAR
ENDED JUNE 30, 1919.

Ordinary Repairs to Canals — Erie Canal
Chapter 151, Laws of 1918

NAME	Rank	Rate of compensation	* Services	Travel	Total
L. C. Hulburd.........	Senior assistant engineer......	$3,540 per year	$1,239 00	$100 31	$1,339 31
R. S. Greenman.......	Senior assistant engineer......	3,300 per year	2,365 00	39 72	2,404 72
T. L. Watkins.........	Assistant engineer.............	2,580 per year	473 00	....	473 00
C. D. Burrus..........	Junior assistant engineer......	1,800 per year	235 71	....	235 71
Parkes D. Wendell.....	Estimate clerk................	3,000 per year	1,100 00	....	1,100 00
Hattie A. Dell.........	Stenographer..................	1,200 per year	880 00	....	880 00
			$6,292 71	$140 03	$6,432 74
	Incidental Expenses				
Stationery and printing..				$17 65	
Miscellaneous..				126 85	144 50
Total...					$6,577 24

Ordinary Repairs to Canals — Champlain Canal
Chapter 151, Laws of 1918

NAME	Rank	Rate of compensation	* Services	Travel	Total
L. C. Hulburd.........	Senior assistant engineer........	$3,540 per year	$619 50	$84 05	$703 55
R. S. Greenman.......	Senior assistant engineer.......	3,300 per year	1,182 50	11 41	1,193 91
T. L. Watkins.........	Assistant engineer.............	2,580 per year	172 79	....	172 79
C. D. Burrus..........	Junior assistant engineer......	1,800 per year	82 50	....	82 50
Parkes D. Wendell.....	Estimate clerk................	3,000 per year	550 00	....	550 00
Hattie A. Dell.........	Stenographer..................	1,200 per year	440 00	....	440 00
			$3,047 29	$95 46	$3,142 75
	Incidental Expenses				
Stationery and printing..				$6 95	
Postage...				4 27	
Miscellaneous..				268 79	280 01
Total...					$3,422 76

* Includes additional compensation of 10 per cent allowed above base rate.

Construction of Barge Canal — Head Office Account

Chapter 147, Laws of 1903, and amendatory laws

NAME	Rank	Rate of compensation	* Services	Travel	Total
D. B. La Du	Special deputy state engineer...	$7,000 per year	$291 67		$291 67
F. P. Williams	Special deputy state engineer...	7,000 per year	1,750 04	$167 76	1,917 80
W. B. Landreth	Deputy state engineer	5,000 per year		29 34	29 34
H. C. Allen	Consulting engineer	60 00 per day	720 00	101 26	821 26
Henry Goldmark	Consulting engineer	60 00 per day	720 00	156 97	876 97
E. E. Haskell	Consulting engineer	60 00 per day	600 00	119 06	719 06
W. B. Landreth	Consulting engineer	6,000 per year	3,301 62	120 77	3,422 39
E. C. Moore	Consulting engineer	60 00 per day	720 00	88 07	808 07
C. H. Paul	Consulting engineer	100 00 per day	400 00	155 07	1,555 07
Joseph Ripley	Consulting engineer	7,200 per year	406 45		406 45
R. E. Phillips	Engineer of claims	4,800 per year	542 50	874 66	5,417 16
Fred J. Wagner	Engineering expert	4,000 per year	591 39	250 88	842 27
C. C. Egbert	Expert in electrical design	20 00 per day	40 00	9 50	49 50
H. D. Alexander	Senior assistant engineer	3,540 per year	1,298 00	118 94	1,416 94
D. H. Daley	Senior assistant engineer	3,300 per year	3,164 77	25 57	3,190 34
R. S. Greenman	Senior assistant engineer	3,300 per year		856 72	856 72
C. H. MacCulloch	Senior assistant engineer	3,540 per year	2,315 60	23 11	2,338 71
J. M. C. Quarles de Quarles	Senior assistant engineer	3,300 per year	3,300 50		3,300 50
G. G. Underhill	Senior assistant engineer	3,540 per year	3,558 50	383 85	3,942 35
N. E. Whitford	Senior assistant engineer	3,300 per year	3,547 50	213 98	3,761 48
C. H. Wood	Senior assistant engineer	3,300 per year	241 59		241 59
G. W. Codwise	Confidential assistant	4,400 per year	780 64	113 90	894 54
Fred J. Wagner	Confidential assistant	4,000 per year	1,208 23	174 61	1,382 84
L. D. McCormac	Private secretary	2,400 per year		178 16	178 16
C. R. Waters	Private secretary	2,340 per year		10 98	10 98
J. J. Allen	Canal clerk	1,800 per year	1,935 00		1,935 00
C. B. Dunham, Jr	Clerk	2,400 per year	2,425 62		2,425 62
J. T. Gorman	Clerk	1,980 per year	1,943 25	138 58	2,081 83
J. C. Guffin	Clerk	1,800 per year	1,749 75		1,749 75
J. E. F Minnock	Clerk	1,980 per year	1,943 25		1,943 25
G. T. Waterman	Clerk	1,200 per year	1,210 00		1,210 00
W. E. Brower	Messenger	720 per year	353 24		353 24
F. J. Murphy	Page	480 per year	90 93		90 93
Mary Broughton	Stenographer	900 per year	940 50		940 50
Nelle Clark	Stenographer	1,200 per year	1,320 00		1,320 00
W. L. Collins	Stenographer	1,200 per year	550 00		550 00
Agnes Fogerty	Stenographer	1,200 per year	1,320 00		1,320 00
Mary G. Harrington	Stenographer	1,200 per year	933 23		933 23
Grace Haswell	Stenographer	1,350 per year	1,485 00		1,485 00
Anna W. Newton	Stenographer	1,350 per year	1,485 00		1,485 00
J. J. Tobin	Stenographer	1,800 per year	897 63		897 63
Jessie J. Weller	Stenographer	900 per year	781 00		781 00
T. S. Bailey	Assistant engineer	2,580 per year	2,537 00		2,537 00
J. C. Bell	Assistant engineer	2,580 per year	55 48		55 48
H. W. Benedict	Assistant engineer	2,580 per year	631 55		631 55
F. E. Blake	Assistant engineer	2,580 per year	1,710 87		1,710 87
H. E. Brainard	Assistant engineer	2,580 per year	49 00		49 00
Clark Brown	Assistant engineer	2,208 per year	2,171 20		2,171 20
N. E. Cottrell	Assistant engineer	1,980 per year	735 05		735 05
C. R. De Graff	Assistant engineer	2,580 per year	185 82	6 88	192 70
G. E. Gibson	Assistant engineer	2,580 per year	1,655 50		1,655 50
M. W. Grimes	Assistant engineer	2,160 per year	2,124 00	160 74	2,284 74
F. B. Hall	Assistant engineer	2,580 per year	1,695 64	18 10	1,713 74
F. W. Harris	Assistant engineer	2,580 per year	91 55		91 55
A. G. Hayden	Assistant engineer	2,580 per year	403 88		403 88
T. R. Hazelum	Assistant engineer	2,160 per year	181 50		181 50
G. D. Kellogg	Assistant engineer	2,580 per year	862 77		862 77
H. C. Kline	Assistant engineer	2,580 per year	126 70		126 70
J. B. Maguire	Assistant engineer	2,340 per year	1,527 61	14 02	1,541 63
W. S. McDowell	Assistant engineer	2,580 per year	1,453 57		1,453 57
R. H. Merrill	Assistant engineer	2,580 per year	374 52		374 52
C. W. Morris, Jr	Assistant engineer	2,160 per year	1,071 57		1,071 57
J. T. Murphy	Assistant engineer	2,340 per year	856 70		856 70
E. P. Neuschwander	Assistant engineer	2,580 per year	2,773 50	99 89	2,873 39
J. A. O'Donnell	Assistant engineer	2,160 per year	2,002 64		2,002 64
C. E. Quimby	Assistant engineer	2,160 per year	270 60		270 60
E. G. Raynor	Assistant engineer	2,580 per year	187 26		187 26

* Includes additional compensation of 10 per cent allowed above base rate.

Construction of Barge Canal — Head Office Account — (Continued)

Chapter 147, Laws of 1903, and amendatory laws

NAME	Rank	Rate of compensation	* Services	Travel	Total
L S Rickard	Assistant engineer	$2,160 per year	$1,334 55	...	$1,334 55
H. J. Scheuermann	Assistant engineer	2,580 per year	1,318 74	$10 28	1,329 02
E. G. Semon	Assistant engineer	2,340 per year	125 81	.....	125 81
H. S. Sparr	Assistant engineer	2,160 per year	2,124 00	.....	2,124 00
S. R. Tighe	Assistant engineer	1,980 per year	209 00	...	209 00
C. R. Waters	Assistant engineer	2,340 per year	463 56	..	463 56
T. L. Watkins	Assistant engineer	2,580 per year	.....	17 45	17 45
L. C. West	Assistant engineer	2,580 per year	239 41	19 62	259 03
J. M. Angus	Junior assistant engineer	1,800 per year	486 13	..	486 13
Leroy Bamer	Junior assistant engineer	1,560 per year	41 52		41 52
J. F. Blaise	Junior assistant engineer	1,800 per year	132 50	.....	132 50
C. D. Burrus	Junior assistant engineer	1,800 per year	945 00	.....	945 00
H. F. Eagan	Junior assistant engineer	1,800 per year	163 94	.....	163 94
B. Gasier	Junior assistant engineer	1,320 per year	121 00	...	121 00
C J Grace, Jr.	Junior assistant engineer	1,200 per year	241 29	.....	241 29
A E. Green	Junior assistant engineer	1,800 per year	312 79	.....	312 79
J S. Heath	Junior assistant engineer	1,320 per year	565 97		565 97
W J. Henk	Junior assistant engineer	1,800 per year	1,643 28		1,643 28
J. S. Hyman	Junior assistant engineer	1,800 per year	377 13	.....	377 13
G. B. Kelley	Junior assistant engineer	1,800 per year	1,811 50		1,811 50
C. T. Kniskern	Junior assistant engineer	1,800 per year	945 00	.....	945 00
Jacob Labishiner	Junior assistant engineer	1,320 per year	335 50		335 50
G D. Meer	Junior assistant engineer	1,800 per year	1,852 50	51 34	1,903 84
Charles Messina	Junior assistant engineer	1,560 per year	66 73		66 73
P. R. Murray	Junior assistant engineer	1,800 per year	1,935 00	.....	1,935 00
Mott Palmer	Junior assistant engineer	1,800 per year	15 97		15 97
G L Schillner	Junior assistant engineer	1,800 per year	1,151 50	.....	1,151 50
Paul Scully	Junior assistant engineer	1,320 per year	1,243 96		1,243 96
R. B Smith	Junior assistant engineer	1,800 per year	245 54	.....	1,245 54
G G. Sweet	Junior assistant engineer	1,560 per year	534 00	.....	1,534 00
L. E. Turpit	Junior assistant engineer	1,560 per year	449 33	.....	449 33
Frank Van Zile	Junior assistant engineer	1,680 per year	1,421 07	.....	1,421 07
L. B. Westfall	Junior assistant engineer	1,800 per year	1,892 42	21 88	1,914 30
C. P. Wiweke	Junior assistant engineer	1,800 per year	772 38	.....	772 38
M. E Baker	Engineering assistant	1,080 per year	1,171 50	.....	1,171 50
J. F. Duffy	Engineering assistant	1,080 per year	807 96	.....	807 96
A. E. Halligan	Engineering assistant	1,080 per year	306 05	.....	306 05
C. L. Hawkins	Engineering assistant	960 per year	136 96	.....	136 96
William Leffler	Engineering assistant	960 per year	330 00	.....	330 00
C. W. Wood	Engineering assistant	840 per year	35 93	.....	35 93
E. V. Allendorph	Inspector of engineering works	1,560 per year	1,677 00	.....	1,677 00
M. S. Bierce	Inspector of engineering works	1,560 per year	1,534 00	.....	1,534 00
F. B. Kraft	Inspector of engineering works	60 per year	1,677 00	.....	1,677 00
E. H. Wetsel	Inspector of engineering works	60 per year	1,677 00	.....	1,677 00
J. Ellis Hurrle	Boatman	3 00 per day	359 70	19 71	379 41
W. J. Atkinson	Laborer	50 per day	1,003 75	.....	1,003 75
John Cullen	Laborer	50 per day	860 75	.....	860 75
James Daly	Laborer	50 per day	264 00	.....	264 00
John Dooley	Laborer	50 per day	792 00	.....	792 00
William Felstead	Laborer	50 per day	860 75	.....	860 75
Joseph Habbinger	Laborer	50 per day	816 75	.....	816 75
David Horner	Laborer	50 per day	896 50	.....	896 50
J. M. Macdonald	Laborer	50 per day	860 75	.....	860 75
G. F. Marcoux	Laborer	50 per day	41 25	.....	41 25
Filadelfo Mondello	Laborer	50 per day	860 75	.....	860 75
T. Rattoone	Laborer	50 per day	811 25	.....	811 25
J. W. Shook	Laborer	50 per day	126 50	.....	126 50
W. J. Smith	Laborer	50 per day	1,003 75	.....	1,003 75
H. J. Soule	Laborer	50 per day	860 75	.....	860 75
Henry Strobel	Laborer	50 per day	860 75	.....	860 75
M. J. Tanner	Laborer	50 per day	863 50	.....	863 50
J. R. Van Schoonhoven	Laborer	50 per day	860 75	.. ..	860 75
E. Van Truen	Laborer	50 per day	860 75	.....	860 75
F. E. Davis	Chauffeur	1,650 per year	1,031 25	49 23	1,080 48
J. J. Finn	Chauffeur	1,650 per year	1 886 84	47 30	934 14
H. W. Nutter	Chauffeur	1,650 per year	1,773 75	96 35	1,870 10
F. M. Hill	Title maker	1,560 per year	1,534 00	.....	1,534 00
E. M. Chamberlain	Night watchman	960 per year	172 00	.....	172 00

* Includes additional compensation of 10 per cent allowed above base rate.

Construction of Barge Canal — Head Office Account — (Conc'uded)

Chapter 147, Laws of 1903, and amendatory laws

NAME	Rank	Rate of compensation	* Services	Travel	Total
Theresa M. Stubbing..	Telephone operator.. ..	$840 per year	$924 00	....	$924 00
Catherine Ryan .	Charwoman	480 per year	528 00	....	528 00
Lybrand, Ross Bros. & Montgomery	Public accountants . .	....	37,828 59	$7,696 63	45,525 22

			$177,012 68	$12,641 16	$189,653 84

Incidental Expenses

Instruments, tools and appliances..	$920 89
Office rent	3,643 27
Fuel and light	45 84
Stationery and printing	3,949 10
Postage	1,544 56
Telephone and telegraph...	1,860 54
Miscellaneous	11,348 75

		23,312 95
Total	....	$212,966 79

Construction of Barge Canal — Erie Canal

Chapter 147, Laws of 1903, and amendatory laws

NAME	Rank	Rate of compensation	* Services	Travel	Total
E. A. Lamb....	Senior assistant engineer........	$3,300 per year	$3,245 00	$412 35	$3,657 35
A. E. Steere.	Senior assistant engineer........	3,300 per year	39 03	43 41	82 44
Parkes D. Wendell	Estimate clerk .	3,000 per year	1,575 00		1,575 00
G. P. Gleason.........	Stenographer to division engineer	2,100 per year	927 50	. .	927 50
Edna M. Pickert	Stenographer	1,200 per year	1,210 00		1,210 00
M. J. Sullivan...	Stenographer	1,200 per year	1,320 00		1,320 00
C. A. Curtis. .	Assistant engineer	2,580 per year		32 95	32 95
C. R. De Graff	Assistant engineer ..	2,580 per year	714 41	.	714 41
G. A. Ensign	Assistant engineer	2,160 per year	893 89	130 00	1,023 89
R. D. Hayes..........	Assistant engineer	2,580 per year		13 05	13 05
M. E. James	Assistant engineer........	2,580 per year	2,773 50	996 34	3,769 84
B. T. Kenyon......	Assistant engineer......	2,580 per year	353 25	29 77	383 02
T. J. Looney........	Assistant engineer	1,980 per year	961 56	....	961 56
A. P. Mussi.	Assistant engineer........	2,340 per year	1,290 01	719 69	2,009 70
C. G. Ranney	Assistant engineer	2,580 per year	2,773 50	314 79	3,088 29
T. L. Watkins........	Assistant engineer	2,580 per year	1,203 21	.	1,203 21
Leroy Bailey... .	Junior assistant engineer .	1,200 per year	638 71	.	638 71
A. H. Charchian....	Junior assistant engineer	1,320 per year	626 74	....	626 74
E. F. Dossert.......	Junior assistant engineer ..	1,680 per year	27 10	....	27 10
H. F. Eagan........	Junior assistant engineer ..	1,800 per year	774 96	....	774 96
F. B. Gifford.........	Junior assistant engineer	1,440 per year	810 45	....	810 45
F. E. Gillen	Junior assistant engineer	1,800 per year	1,849 66		1,849 66
W. M. Griffith .	Junior assistant engineer	1,800 per year	1,935 00	395 42	2,330 42
H. W. Jewell...	Junior assistant engineer	1,800 per year	1,312 52	346 18	1,658 70
C. T. McLean .	Junior assistant engineer ..	1,440 per year	517 96	.	517 96
William Mangan......	Junior assistant engineer .	1,560 per year	983 93	.	983 93
M. J. Quinn..........	Junior assistant engineer .	1,440 per year	798 48	96 41	894 89
R. B. Smith..........	Junior assistant engineer ..	1,800 per year	460 58	..	460 58
C. E. Vedder........	Junior assistant engineer .	1,440 per year	304 50	....	304 50
C. A. Wilbur........	Junior assistant engineer ..	1,800 per year	1,274 35	190 85	1,465 20
F. S. Belotti ..	Engineering assistant............	1,080 per year	1,073 03	..	1,073 03
E. E. Fobes .	Engineering assistant.	840 per year	76 25	....	76 25
D. A. Gillette .	Engineering assistant.	840 per year	179 66	. . .	179 66
George Hinds. .	Engineering assistant.	840 per year	89 42	..	89 42
J. C. Quinterro........	Engineering assistant........	900 per year	685 69	....	685 69

* Includes additional compensation of 10 per cent allowed above base rate.

Construction of Barge Canal — Erie Canal — (Continued)

Chapter 147, Laws of 1903, and amendatory laws

NAME	Rank	Rate of compensation	* Services	Travel	Total
F. B. Stoddard	Engineering assistant.	$1,080 per year	$710 14	......	$710 14
C. B. Tebo	Engineering assistant	1,080 per year	1,168 20	....	1,168 20
T. M. Oliver	Inspector of engineering works...	1,560 per year	1,529 39	....	1,529 39
Raymond Comrie......	Boatman......................	3 00 per day	660 00	....	660 00
Harold Folmsbee......	Boatman......................	3 00 per day	1,023 00	....	1,023 00
H. D. Horning........	Boatman......................	3 00 per day	108 90	....	108 90
A. B. Starin..	Boatman......................	3 00 per day	1,065 90	....	1,065 90
H. L. Crouse..	Laborer.......................	2 50 per day	495 00	....	495 00
Thomas Dalton.	Laborer	2 50 per day	52 25	....	52 25
William De Forest....	Laborer	2 50 per day	514 25	.	514 25
R. N. Kay............	Laborer.......................	2 50 per day	19 25	....	19 25
John Lavery..........	Laborer..	2 50 per day	85 25	....	85 25
T. F. Madden........	Laborer.......................	2 50 per day	940 50	....	940 50
M. Mets..... ..	Laborer.......................	2 50 per day	567 87	....	567 87
P. Sajta.............	Laborer.......................	2 50 per day	522 50	....	522 50
H. J. Richardson. . .	Photographer	1,680 per year	1,008 00	$43 27	1,051 27
Harry Bishton	Gage reader	14 00 per month	154 00	....	154 00
Guy Bracebridge ...	Gage reader	5 00 per month	55 00	....	55 00
Carlton Cornwell .	Gage reader	7 00 per month	35 00	....	35 00
Forrest Devle. ...	Gage reader	7 00 per month	63 00	....	63 00
H. C. Dowling... .	Gage reader	7 00 per month	77 00	....	77 00
P. C. Earl............	Gage reader	7 00 per month	84 00	....	84 00
L. E. Jeffords. .	Gage reader	10 00 per month	36 13	....	36 13
Lloyd Kast....	Gage reader	7 00 per month	84 00	....	84 00
Richard Kilmartin	Gage reader	7 00 per month	84 00	....	84 00
Clark Kyser... ...	Gage reader...................	7 00 per month	77 00	....	77 00
Peter Lebeis. .	Gage reader...................	7 00 per month	56 00	....	56 00
Oscar Lockwood....	Gage reader	7 00 per month	84 00	....	84 00
C. F. Loring....	Gage reader	7 00 per month	14 00	....	14 00
James Murphy ..	Gage reader	9 00 per month	18 00	....	18 00
Rose Murphy.........	Gage reader...	9 00 per month	45 00	....	45 00
Fred Pentland......	Gage reader . . .	9 00 per month	55 16	....	55 16
P. C. Pickard.........	Gage reader	7 00 per month	49 00	....	49 00
Antoine Plouffe.	Gage reader . . .	9 00 per month	36 00	....	36 00
J. Reepmeyer........	Gage reader .	10 00 per month	120 00	....	120 00
William Reihl .	Gage reader	9 00 per month	52 84	....	52 84
A. W. Spencer........	Gage reader . . .	7 00 per month	49 00	..	49 00
Minnie E. White . ..	Gage reader . . .	7 00 per month	42 00	....	42 00
Robert Wilson	Gage reader	10 00 per month	10 00	. . . . .	10 00
C. E. Wing...........	Gage reader	10 00 per month	110 00	....	110 00
C. W. Young	Gage reader . . .	14 00 per month	140 00	....	140 00

                 $45,799 38 $3,764 48 $49,563 86

Incidental Expenses

Office rent.	$1,030 00	
Fuel and light.	141 17	
Stationery and printing	9 78	
Postage.	79 69	
Telephone and telegraph.	408 86	
Miscellaneous.	1,003 54	
		2,673 04
Total		$52,236 90

* Includes additional compensation of 10 per cent allowed above base rate.

Construction of Barge Canal — Champlain Canal
Chapter 147, Laws of 1903, and amendatory laws

NAME	Rank	Rate of compensation	* Services	Travel	Total
John Schade	Cashier	$2,100 per year	$2,195 10	. .	$2,195 10
J. E. Phinney	Stenographer	1,320 per year	1,452 00		1,452 00
C. A. Curtis	Assistant engineer	2,580 per year	2,270 68	$735 20	3,005 88
J. B. Foote	Assistant engineer	2,340 per year	2,301 00	1,114 73	3,415 73
R. G. Gibson	Assistant engineer	2,160 per year	1,704 00	7 85	1,711 85
F. W. Harris	Assistant engineer	2,580 per year	1,327 45		1,327 45
R. D. Hayes	Assistant engineer	2,580 per year	2,773 50	1,151 39	3,924 89
Harold Bristol	Junior assistant engineer	1,200 per year	1,320 00		1,320 00
D. E. Damon	Junior assistant engineer	1,800 per year	1,935 00		1,935 00
G. E. Deutschbein	Junior assistant engineer	1,800 per year	1,935 00	83 65	2,018 65
Mott Palmer	Junior assistant engineer	1,800 per year	1,280 00	42 20	1,322 20
J. P. Walsh	Junior assistant engineer	1,680 per year	1,592 62		1,592 62
G. A. Rogers	Engineering assistant	1,080 per year	22 35		22 35
Thomas Ryan, Jr	Engineering assistant	1,080 per year	95 61		95 61
J. C. Leyland	Boatman	3 00 per day	42 90		42 90
G. E. McElroy	Boatman	3 00 per day	283 80		283 80
N. H. McHerd	Boatman	3 00 per day	1,204 50	. . .	1,204 50
Harry McMahon	Boatman	3 00 per day	435 60		435 60
Ephraim Newland	Boatman	3 00 per day	521 40		521 40
Edward Ryan	Boatman	3 00 per day	528 00		528 00
F. H. Crandall	Laborer	2 50 per day	847 00	. . .	847 00
B. K. Ellis	Laborer	2 50 per day	871 20		871 20
F. J. Hilfinger	Laborer	2 50 per day	220 00		220 00
J. F. Malin	Laborer	2 50 per day	506 00	.	506 00
L. Saunders	Laborer	2 50 per day	126 50		126 50
H. J. Richardson	Photographer	1,680 per year	560 00	57 41	617 41
B. H. Bennett	Gage reader	7 00 per month	3 50		3 50
E. H. Bowker	Gage reader	7 00 per month	49 00	. .	49 00
Charles Cheney	Gage reader	7 00 per month	7 00		7 00
J. H. Donnelly	Gage reader	7 00 per month	84 00	.	84 00
W. B. Dunstan	Gage reader	7 00 per month	87 50		87 50
G. E. Fifield	Gage reader	7 00 per month	84 00		84 00
A. B. Fisher	Gage reader	7 00 per month	56 00		56 00
C. F. Mayhew	Gage reader	8 00 per month	88 00		88 00
J. T. Morris	Gage reader	8 00 per month	88 00	. . . .	88 00
Henry Palmer	Gage reader	7 00 per month	77 00	. . . .	77 00
Byron Stedman	Gage reader	7 00 per month	10 38	.	10 38
F. N. Wells	Gage reader	7 00 per month	21 00	. . . .	21 00
			$29,006 59	$3,192 43	$32,199 02

Incidental Expenses

Office rent	$574 00	
Fuel and light	102 22	
Stationery and printing	11 44	
Postage	129 32	
Telephone and telegraph	246 37	
Miscellaneous	753 39	
		1,816 74
Total		$34,015 76

* Includes additional compensation of 10 per cent allowed above base rate.

Construction of Barge Canal Terminals

Chapter 746, Laws of 1911, and amendatory laws

NAME	Rank	Rate of compensation	* Services	Travel	Total
D. B. La Du	Special deputy state engineer...	$7,000 per year	$291	$4 00	$295 66
F. P. Williams	Special deputy state engineer...	7,000 per year	1,749	211 07	1,961 03
H. D. Alexander	Senior assistant engineer	3,540 per year	649		649 00
Edward Anderberg	Senior assistant engineer	3,540 per year	2,729	88 94	2,818 20
A. W. Conner	Senior assistant engineer	3,300 per year	3,300	75 03	3,375 53
C. H. MacCulloch	Senior assistant engineer	3,540 per year	1,489	115 46	1,605 36
C. H. Wood	Senior assistant engineer	3,300 per year	2,458	70 86	2,529 72
W. B. Landreth	Consulting engineer	6,000 per year	35	74 05	109 53
E. P. Goodrich	Expert on terminal construction	40 00 per day	120	23 74	143 74
H. McL. Harding	Expert on terminal construction	40 00 per day	40 66	27 05	67 05
M. W. Williams	Expert on terminal construction	40 00 per day	1,000 00	139 41	1,139 41
F. J. Wagner	Confidential assistant	4,000 year		95 61	95 61
F. J. Wagner	Engineering expert	4,000 year	1,100 00	184 18	1,284 18
C. R. Waters	Private secretary	2,340 per year		98 41	98 41
L. D. McCormac	Private secretary	2,400 per year		122 40	122 40
W. S. Ryan	Confidential clerk and stenographer	2,400 per year		30 45	30 45
Fred C. Stahl	Bookkeeper	1,800 year			325 97
James E. Stewart	Clerk	1,980 year	1,		1,778 25
G. P. Gleason	Stenographer to division engineer	2,100 year	1		1,330 00
Emily P Hofmann	Stenographer	900 year			863 50
S. R. Bellows	Assistant engineer	2,580 per year	366		90 16
H. W. Benedict	Assistant engineer	2,580 per year	214 66		214 50
F. E. Blake	Assistant engineer	2,580 er year	1,062 63		1,062 63
W. C. Bratton	Assistant engineer	2,580 p r year	2,	80 18	2,853 68
P. H. Budd	Assistant engineer	2,208 per year	773	20	808 49
W. L. Calcr	Assistant engineer	2,580 per year	2,363	473 44	3,246 94
Horace Corbin	Assistant engineer	2,580 per year	2,626	2 97	2,629 94
N. E. Cottrell	Assistant engineer	1,980 per year	1,393		1,393 45
C. A. Curtis	Assistant engineer	2,580 per year	130	53 28	183 67
J. B. Doughty	Assistant engineer	2,340 per year	2,	54 15	2,384 40
G A. Ensign	Assistant engineer	2,160 per year			351 91
Ely Gamse	Assistant engineer	2,580 per year	2,	99 31	2,872 81
R. G. Gibson	Assistant engineer	2,160 per year		12 58	36 58
H. W. Hale	Assistant engineer	2,580 per year	1,	105 56	1,459 41
F. B. Hall	Assistant engineer	2,580 per year	330 50	31 55	847 85
F W. Earris	Assistant engineer	2,580 per year	566 66	30 33	503 33
A. G. Hayden	Assistant engineer	2,580 per year	2,133 12		2,133 12
R. D. Hayes	Assistant engineer	2,580 per year	236 50	9 50	246 00
L. T. Howard	Assistant engineer	2,580 per year	2,556 84	106 28	2,663 12
G. D. Kellogg	Assistant engineer	2,580 per year	157 67		157 67
H. C. Kline	Assistant engineer	2,580 per year	610 32	31 62	641 94
F. T. Lawton	Assistant engineer	2,580 per year	2,526 50	62 55	2,589 05
T. J. Loonie	Assistant engineer	1,980 per year	647 02		647 02
W. S. McDowel	Assistant engineer	2,580 per year	277 18		277 18
J. B. Maguire	Assistant engineer	2,340 per year	773 39		773 39
C W Morris. Jr	Assistant engineer	2,160 per year	1,052 43		1,052 43
A. P. Mussi	Assistant engineer	2,340 per year	1,225 49	111 45	1,336 94
E. C. Olcott	Assistant engineer	2,208 per year	2,157 92		2,157 92
C E. Quimby	Assistant engineer	2,160 per year	1,698 41		1,698 41
E. G. Raynor	Assistant engineer	2,580 per year	1,106 21		1,106 21
A C. Richards	Assistant engineer	2,580 per year	2,773 50		2,773 50
H. J. Scheuermann	Assistant engineer	2,580 per year	1,454 76	2 50	1,457 26
Rupert Sturtevant	Assistant engineer	2,580 per year	2,422 56	194 72	2,617 28
S. R. Tighe	Assistant engineer	1,980 per year	165 00		165 00
C. R. Waters	Assistant engineer	2,340 per year	607 20		607 20
T. L. Watkins	Assistant engineer	2,580 per year	924 50	28 97	953 47
L. C. West	Assistant engineer	2,580 per year	2,287 09		2,287 09
J. M. Angus	Junior assistant engineer	1,800 per year	546 45		546 45
H. T. Arnold	Junior assistant engineer	1,800 per year	1,706 82		1,706 82
J. F. Blaise	Junior assistant engineer	1,800 per year	1,637 50		1,637 50
J. J. Carroll	Junior assistant engineer	1,440 per year	39 03		39 03
A. H. Charchian	Junior assistant engineer	1,320 per year	72 80		72 80
J. A. Daley	Junior assistant engineer	1,440 per year	1,454 00		1,454 00
L. A. Denner	Junior assistant engineer	1,440 per year	11 61		11 61
John Edelstein	Junior assistant engineer	1,680 per year	1,621 21		1,621 21
B. Gaxier	Junior assistant engineer	1,320 per year	1,163 16		1,163 16
A. E. Green	Junior assistant engineer	1,800 per year	167 84		167 84
Leroy Greenalch	Junior assistant engineer	1,800 per year	373 82		373 82

* Includes additional compensation of 10 per cent allowed above base rate.

Construction of Barge Canal Terminals — (Continued)

Chapter 746, Laws of 1911, and amendatory laws

NAME	Rank	Rate of compensation	* Services	Travel	Total
J. E. Hall	Junior assistant engineer	$1,800 per year	$1,935 00		$1,935 00
J. S. Heath	Junior assistant engineer	1,320 per year	765 03		765 03
H. W. Henderson	Junior assistant engineer	1,440 per year	70 83		70 83
W. J. Henk	Junior assistant engineer	1,800 per year	168 22		168 22
J. A. Husband	Junior assistant engineer	1,320 per year	318 88		318 88
J. S. Hyman	Junior assistant engineer	1,800 per year	214 51		214 51
Samuel Jaffe	Junior assistant engineer	1,200 per year	275 00		275 00
H. W. Jewell	Junior assistant engineer	1,800 per year	409 75	$59 28	469 03
H. Kabak	Junior assistant engineer	1,320 per year	1,075 34		1,075 34
C. F. Keale, Jr	Junior assistant engineer	,440 per year	503 97		503 97
H. C. Kelly	Junior assistant engineer	,680 per year	47 67		47 67
S. Levine	Junior assistant engineer	,800 per year	657 45		657 45
William Mangan	Junior assistant engineer	,560 per year	119 91		119 91
Charles Messina	Junior assistant engineer	,560 per year	47 67		47 67
Charles Montag	Junior assistant engineer	,800 per year	384 64		384 64
D. C. Ogsbury	Junior assistant engineer	,800 per year	1,935 00		1,935 00
Mott Palmer	Junior assistant engineer	,800 per year	10 00	10 80	20 80
M. J. Quinn	Junior assistant engineer	1,440 per year		13 00	13 00
G. L. Schillner	Junior assistant engineer	,800 per year	660 00		660 00
Isie Spahn	Junior assistant engineer	1,320 per year	1,111 81		1,111 81
Isaac Stern	Junior assistant engineer	,440 per year	1,402 74		1,402 74
L. E. Turpit	Junior assistant engineer	,560 per year	250 00		250 00
F. M. Van Zile	Junior assistant engineer	,680 per year	230 93		230 93
C. A. Wilbur	Junior assistant engineer	,800 per year	64 52		64 52
C. P. Wiweke	Junior assistant engineer	,800 per year	1,162 62		1,162 62
F. S. Belotti	Engineering assistant	,080 per year	15 97		15 97
J. F. Duffy	Engineering assistant	,080 per year	332 14		332 14
W. R. Glock	Engineering assistant	900 per year	209 00		209 00
A. E. Halligan	Engineering assistant	1,080 per year	881 95		881 95
A. Heckman, Jr	Engineering assistant	840 per year	161 52		161 52
G. Hinds	Engineering assistant	840 per year	4 97		4 97
Joseph Hoehlin	Engineering assistant	840 per year	278 11		278 11
William Leffler	Engineering assistant	960 per year	118 87		118 87
G. W. Nostrand	Engineering assistant	900 per year	336 16		336 16
J. C. Quinterro	Engineering assistant	900 per year	73 31		73 31
J. J. Raup	Engineering assistant	1,080 per year	1,178 42		1,178 42
G. A. Rogers	Engineering assistant	1,080 per year	198 00		198 00
F. B. Stoddard	Engineering assistant	,080 per year	477 86		477 86
E. A. Terrell	Engineering assistant	840 per year	46 20		46 20
George Terwilliger	Engineering assistant	1,080 per year	1,188 00		1,188 00
T. J. Torpy, Jr	Engineering assistant	900 per year	43 63		43 63
W. H. H. Klinkhart	Inspector of engineering works	1,560 per year	1,677 00		1,677 00
H. Kramer	Inspector of engineering works	1,560 per year	1,677 00		1,677 00
T. M. Oliver	Inspector of engineering works	,560 per year	4 61		4 61
George Alexander	Boatman	3 00 per day	224 40		224 40
A. A. Boles	Boatman	3 00 per day	1,197 90		1,197 90
Louis Cicio	Boatman	3 00 per day	709 50		709 50
Raymond Comrie	Boatman	3 00 per day	151 80		151 80
Walter Cormany	Boatman	3 00 per day	1,204 50		1,204 50
B. K. Ellis	Boatman	3 00 per day	52 80		52 80
P. F. Fitzgerald	Boatman	3 00 per day	1,042 80		1,042 80
Harold Folmsbee	Boatman	3 00 per day	9 90		9 90
H. D. Horning	Boatman	3 00 per day	429 00		429 00
J. A. Jacobson	Boatman	3 00 per day	1,059 30		1,059 30
J. J. Kelly	Boatman	3 00 per day	1,062 60		1,062 60
Giles D. Long	Boatman	3 00 per day	128 70		128 70
Harry McMahon	Boatman	3 00 per day	3 30		3 30
Ephraim Newland	Boatman	3 00 per day	3 30		3 30
E. W. Reilly	Boatman	3 00 per day	1,204 50		1,204 50
George Reuling	Boatman	3 00 per day	554 40		554 40
W. B. Rowland	Boatman	3 00 per day	844 80		844 80
M. J. Sims	Boatman	3 00 per day	1,194 60		1,194 60
J. W. Turner	Boatman	3 00 per day	89 10		89 10
H. L. Crouse	Laborer	2 50 per day	11 00		11 00
William De Forest	Laborer	2 50 per day	13 75		13 75
Charles Girard	Laborer	2 50 per day	992 75		992 75
James Hopkins	Laborer	2 50 per day	893 75		893 75
John Lavery	Laborer	2 50 per day	154 00		154 00
Henry Macfarlane	Laborer	2 50 per day		15 40	15 40

* Includes additional compensation of 10 per cent allowed above base rate.

Construction of Barge Canal Terminals — (Concluded)
Chapter 746, Laws of 1911, and amendatory laws

NAME	Rank	Rate of compensation	* Services	Travel	Total
T. F. Madden	Laborer	$2 50 per day	$63 25		$63 25
G. L. Reuling	Laborer	2 50 per day	151 25		151 25
Paul Sajta	Laborer	2 50 per day	8 25		8 25
Lewis Saunders	Laborer	2 50 per day	2 75		2 75
John W. Shook	Laborer	2 50 per day	41 25		41 25
Gilbert Venter	Laborer	2 50 per day	1,003 75		1,003 75
H. J. Richardson	Photographer	1,680 per year	238 00	$132 38	370 38
F. E. Davis	Chauffeur	1,650 per year	742 50	352 26	,094 76
J. J. Finn	Chauffeur	1,650 per year	886 91	282 32	,169 23
H. W. Nutter	Chauffeur	1,650 per year		290 58	290 58
Lybrand, Ross Bros. & Montgomery	Public accountants		18,363 01	3,304 72	21,667 73

			$133,303 97	$7,242 54	$140,546 51

Incidental Expenses

Instruments, tools and appliances	$46 53
Office rent	4,874 97
Fuel and light	190 59
Stationery and printing	2,435 16
Postage	125 61
Telephone and telegraph	621 12
Miscellaneous	6,851 29
	15,145 27
Total	$155,691 78

Hudson River Terminals
Chapter 555, Laws of 1918

NAME	Rank	Rate of compensation	* Services	Travel	Total
W. B. Landreth	Consulting engineer	$6,000 per year	$35 48	$18 82	$54 30
G. W. Codwise	Confidential assistant	4,400 per year		3 18	3 18
B F Cresson, Jr	Expert on terminal construction	per day	160	20 52	180 52
E. P. Goodrich	Expert on terminal construction	day	160	22 55	182 55
H. McL. Harding	Expert on terminal construction	40 00 per day	160	21 95	181 95
M. W. Williams	Expert on terminal construction	40 00 per day	200	42 73	242 73
F. C. Stahl	Bookkeeper	1,800 ear	69		69 03
J. C. Bell	Assistant engineer	2,580 ear	2,308	719 85	3,027 98
G. D. Kellogg	Assistant engineer	2,580 per year	20 00	62 82	83 63
C. T. Kniskern	Junior assistant engineer	1,800 per year	330 00		330 00
William Leffler	Engineering assistant	960 per year	251 40		251 40
Paul Scully	Junior assistant engineer	1,320 per year	36 30		36 30
James Daly	Laborer	2 50 per day	156 75		156 75

			$3,887 90	$912 42	$4,800 32

Incidental Expenses

Postage	$0 37
Miscellaneous	15 89
	16 26
Total	$4,816 58

* Includes additional compensation of 10 per cent allowed above base rate.

Bridge Designers, Engineers, etc.
Chapter 151, Laws of 1918

NAME	Rank	Rate of compensation	* Services	Expenses	Total
H. E. Brainard........	Assistant engineer.............	$2,580 per year	$1,563 50	$25 01	$1,585 51
W. S. McDowell.......	Assistant engineer.............	2,580 per year	236 50	....	236 50
Keuffel & Esser Company.........	....	....	....	128 83	128 83
Postmaster............	....	....	....	49 16	49 16
Total............	....	....	$1,800 00	$300 00	$2,000 00

High Street Bridge, Cohoes
Chapter 181, Laws of 1917, chapter 151, Laws of 1918

NAME	Rank	Rate of compensation	* Services	Travel	Total
C. A. Curtis..........	Assistant engineer.............	$2,580 per year	$372 43	$104 67	$477 10
J. P. Walsh...........	Junior assistant engineer.......	1,680 per year	74 88	....	74 88
B. K. Ellis...........	Laborer...............	2 50 per day	33 00	....	33 00
John Lavery...........	Laborer...............	2 50 per day	220 00	....	220 00
			$700 31	$104 67	$804 98
Incidental Expenses					
Stationery and printing........				$15 68	
Fuel and light..........				0 32	
					16 00
Total...........					$820 98

Schenectady–Scotia Bridge
Chapter 735, Laws of 1917; chapter 634, Laws of 1919

NAME	Rank	Rate of compensation	* Services	Travel	Total
B. A. Davis...........	Consulting engineer...........	$60 00 per day	$300 00	$33 87	$333 87
E. D. Hendricks.......	Senior assistant engineer.......	3,300 per year	419 60	73 04	492 64
C. H. Wood...........	Senior assistant engineer.......	3,300 per year	600 05	....	600 05
G. D. Kellogg.........	Assistant engineer............	2,580 per year	267 02	137 64	404 66
W. S. McDowell.......	Assistant engineer............	2,580 per year	118 25	....	118 25
C. E. Quimby.........	Assistant engineer............	2,160 per year	154 99	....	154 99
J. J. Carroll..........	Junior assistant engineer.......	1,440 per year	138 61	....	138 61
A. H. Charohian.......	Junior assistant engineer.......	1,320 per year	76 11	....	76 11
A. E. Green...........	Junior assistant engineer.......	1,800 per year	115 50	....	115 50
L. Greenalch..........	Junior assistant engineer.......	1,800 per year	227 63	....	227 63
G. D. Meer...........	Junior assistant engineer.......	1,800 per year	82 50	....	82 50
R. B. Smith...........	Junior assistant engineer.......	1,800 per year	172 81	14 44	187 25
William Leffler.......	Engineering assistant.........	960 per year	95 90	....	95 90
C. B. Tebo...........	Engineering assistant.........	1,080 per year	19 80	....	19 80
			$2,788 77	$258 99	$3,047 76
Incidental Expenses					
Miscellaneous.........					308 26
Total...........					$3,356 02

* Includes additional compensation of 10 per cent allowed above base rate.

Sea-wall, Orient – East Marion, L. I.
Chapter 428, Laws of 1918

NAME	Rank	Rate of compensation	* Services	Expenses	Total
Edward Anderberg.....	Senior assistant engineer........	$3,540 per year	$9 76		$9 76
H. W. Benedict	Assistant engineer..............	2,580 per year	37 74	..	37 74
H. W. Hale	Assistant engineer.............	2,580 per year	113 02	$238 74	351 76
F. B. Hall...........	Assistant engineer.............	2,580 per year	69 35		69 35
H. F. Eagan..........	Junior assistant engineer........	1,800 per year	44 71	37 62	82 33
Isie Spahm..........	Junior assistant engineer........	1,320 per year	3 55		3 55
Isaac Stern..........	Junior assistant engineer........	1,440 per year	15 61	..	15 61
J. J. Kelly..........	Boatman....................	3 00 per day	141 90		141 90
Oliver A. Quayle......		....		27 52	27 52
Total.........		....	$435 64	$303 88	$739 52

Blue Line Surveys
Chapter 151, Laws of 1918

NAME	Rank	Rate of compensation	* Services	Travel	Total
G. W. Codwise	Confidential assistant..........	$4,400 per year	$1,484 99	....	$1,484 99
John Schade.......	Cashier.	2,100 per year	62 40	....	62 40
C. R. De Graff......	Assistant engineer......... .	2,580 per year	282 27	....	282 27
G. A. Ensign.......	Assistant engineer..........	2,160 per year	878 20	......	878 20
T. R. Haselum......	Assistant engineer..........	2,160 per year	371 25	......	371 25
Edwin Hilborn......	Assistant engineer..........	2,580 per year	2,633 97	$309 45	2,943 42
O. F. Lewis........	Assistant engineer..........	2,340 per year	1,963 10	....	1,963 10
T. J. Loonie.. ..	Assistant engineer..........	1,980 per year	338 42	....	338 42
Leroy Bailey........	Junior assistant engineer	1,200 per year	393 87	....	393 87
A. H. Charchian.....	Junior assistant engineer.......	1,320 per year	437 13	....	437 13
L. W. Douglas.......	Junior assistant engineer.......	1,200 per year	36 67	....	36 67
Roy Engell.........	Junior assistant engineer.......	1,440 per year	45 83	....	45 83
L. E. Fields........	Junior assistant engineer.......	1,440 per year	284 33	....	284 33
F. B. Gifford........	Junior assistant engineer.......	1,440 per year	504 09	....	504 09
James A. Husband....	Junior assistant engineer.......	1,320 per year	363 89	....	363 89
H. W. Jewell........	Junior assistant engineer.......	1,800 per year	212 73	42 29	255 02
C. T. Kniskern......	Junior assistant engineer .	1,800 per year	330 00	....	330 00
B. A. Krotinger.....	Junior assistant engineer .	1,680 per year	415 48	....	415 48
Jacob Labishiner.....	Junior assistant engineer.......	1,320 per year	74 51	....	74 51
R. B. Smith........	Junior assistant engineer.......	1,800 per year	56 07	....	56 07
F. S. Belotti........	Engineering assistant..........	1,080 per year	99 00	....	99 00
H. H. Glosser.......	Engineering assistant..........	960 per year	268 88	....	268 88
C. L. Hawkins......	Engineering assistant.... .	960 per year	347 48	....	347 48
E. S. Niles, Jr......	Engineering assistant..........	840 per year	4 52	6 49	11 01
J. C. Nolan........	Engineering assistant..........	840 per year	25 67	....	25 67
J. C. Quinterro.....	Engineering assistant..........	900 per year	22 00	....	22 00
Edmund Wilcox......	Engineering assistant..........	840 per year	4 52	$6 49	11 01
F. W. Yates........	Engineering assistant..........	1,080 per year	637 73	4 19	641 92
H. L. Crouse.. ..	Laborer...................	2 50 per day	398 75	....	398 75
W. B. Lounsbury.....	Laborer...................	2 50 per day	211 75	....	211 75
P. L. Mattimore......	Laborer...................	2 50 per day	143 00	....	143 00
M. Mets...........	Laborer...................	2 50 per day	115 50	....	115 50
F. J. Norton........	Laborer...................	2 50 per day	195 25	....	195 25
J. W. Shook........	Laborer...................	2 50 per day	156 75	....	156 75
			$13,800 00	$368 91	$14,168 91

Incidental Expenses

Livery..	$558 50	
Fuel and light...	22 69	
Postage..	33 61	
Office rent..	44 00	
Telephone and telegraph..	22 30	
Miscellaneous...	149 99	
		831 09
Total..		$15,000 00

* Includes additional compensation of 10 per cent allowed above base rate.

Surveys for the State Court of Claims
Chapter 151, Laws of 1918

NAME	Rank	Rate of compensation	* Services	Travel	Total
F. W. Harris..........	Assistant engineer .	$2,580 per year	....	$89 93	$89 93
					$89 93
	Incidental Expenses				
Livery		....		$1 75	
Postage..............		....		0 59	
Miscellaneous............		....		107 73	
					$110 07
Total...................		....			$200 00

Department Surveys
Chapter 151, Laws of 1918

NAME	Rank	Rate of compensation	* Services	Expenses	Total
F. W. Harris	Assistant engineer ..	$2,580 per year	$236 50	$265 10	$501 60
G. D. Kellogg .	Assistant engineer .	2,580 per year	347 23	274 54	621 77
B. T. Kenyon .	Assistant engineer .	2,580 per year	315 33	348 80	664 13
J. J. Carroll ..	Junior assistant engineer	1,440 per year	35 13		35 13
A. H. Charchian .	Junior assistant engineer . ,...	1,320 per year	145 72		145 72
H. F. Eagan .	Junior assistant egnineer	1,800 per year		29 66	29 66
H. W. Henderson	Junior assistant engineer	1,440 per year	81 99	.	81 99
H. C. Kelly .	Junior assistant engineer	1,680 per year	334 40		334 40
C. T. Kniskern	Junior assistant engineer	1,800 per year	31 94		31 94
C. T. MacLean.	Junior assistant engineer	1,440 per year	280 30		280 30
J. A. Waddell	Junior assistant engineer	1,800 per year	179 03		179 03
E. E. Fobes	Engineering assistant	840 per year	177 10		177 10
William Leffler.	Engineering assistant	960 per year	161 45		161 45
W. E. Mullen .	Engineering assistant.	900 per year	49 50		49 50
J. C. Quinterro ...	Engineering assistant	900 per year	48 79		48 79
Thomas Ryan, Jr	Engineering assistant..........	1,080 per year	217 80		217 80
James Daly	Laborer	2 50 per day	16 50		16 50
W. & L. E. Gurley.				19 56	19 56
Keuffel & Esser Co				11 56	11 56
Milton A. Van Hoesen..				20 00	20 00
Total			$2,658 71	$969 22	$3,627 93

State Boundary Line
Chapter 151, Laws of 1918

NAME	Rank	Rate of compensation	* Services	Expenses	Total
H. F. Eagan.. ..	Junior assistant engineer ..	$1,800 per year	$378 26	....	$378 26

* Includes additional compensation of 10 per cent allowed above base rate.

Delaware–Schoharie County Boundary Line
Chapter 559, Laws of 1918

NAME	Rank	Rate of compensation	*Services	Expenses	Total
R. S. Greenman......	Senior assistant engineer........	$3,300 per year	....	$74 24	$74 24
H. C. Kline..........	Assistant engineer.............	2,580 per year	$777 47	797 23	1,574 70
T. Kniskern.....	Junior assistant engineer.	1,800 per year	163 19	......	163 19
	Engineering assistant (provisional)	840 per year	41 07	....	41 07
H. A. Dayton........	Engineering assistant........	840 per year	221 60	....	221 60
F. N. McMaster......	Engineering assistant (provisional)	840 per year	83 55	....	83 55
R. H. Slocum........	Engineering assistant (provisional)	840 per year	126 45	....	126 45
D. B. Hall	Laborer..............	2 50 per day	57 75	....	57 75
Malcolm McPherson...	Laborer..............	2 50 per day	33 00	....	33 00
Total...........			$1,504 08	$871 47	$2,375 55

Saratoga–Warren County Boundary Line
Chapter 561, Laws of 1918

NAME	Rank	Rate of compensation	*Services	Expenses	Total
R. S. Greenman.......	Senior assistant engineer........	$3,300 per year	....	$40 61	$40 61
F. W. Harris.........	Assistant engineer.............	2,580 per year	$645 00	1,072 69	1,717 69
L. E. Turpit.........	Junior assistant engineer.......	1,440 per year	28 31	....	28 31
Dwight Douglas......	Engineering assistant (provisional)	840 per year	156 33	....	156 33
J. L. Loohner........	Engineering assistant (provisional)	840 per year	168 00	....	168 00
M. W. Sarr..........	Engineering assistant (provisional)	840 per year	168 00	....	168 00
C. W. Wood..........	Engineering assistant (provisional)	840 per year	197 86	....	197 86
			$1,363 50	$1,113 30	$2,476 80

Ulster–Greene County Boundary Line
Chapter 562, Laws of 1918; chapter 600, Laws of 1919

NAME	Rank	Rate of compensation	*Services	Expenses	Total
R. S. Greenman.......	Senior assistant engineer........	$3,300 per year	....	$59 32	$59 32
H. C. Kline.........	Assistant engineer.............	2,580 per year	$821 16	1,716 61	2,537 77
H. A. Dayton........	Engineering assistant........	840 per year	219 64	....	219 64
H. F. Eagan.........	Junior assistant engineer........	1,800 per year	118 43	....	118 43
C. W. Wood.........	Engineering assistant (provisional)	840 per year	92 40	....	92 40
R. H. Slocum	Engineering assistant (provisional)	840 per year	50 88	....	50 88
James Daly.........	Laborer..............	2 50 per day	132 00	....	132 00
Elmer Eglinton.......	Laborer..............	2 50 per day	38 50	....	38 50
C. E. George... ..	Laborer..............	2 50 per day	44 00	....	44 00
Earl Gossoo.........	Laborer..............	2 50 per day	123 75	....	123 75
Francis Hagadorn....	Laborer..............	2 50 per day	35 75	....	35 75
Malcolm McPherson...	Laborer..............	2 50 per day	16 50	....	16 50
C H. Satterlee	Laborer..............	2 50 per day	129 25	....	129 25
Roy Satterlee........	Laborer..............	2 50 per day	52 25	....	52 25
Milton A. Van Hoesen.	....		....	32 25	32 25
Total...........			$1,874 51	$1,808 18	$3,682 69

* Includes additional compensation of 10 per cent allowed above base rate.

Land Grants
Chapter 151, Laws of 1918

NAME	Rank	Rate of compensation	* Services	Expenses	Total
E. V. R. Payne	Senior assistant engineer	$3,300 per year	$121 00		$121 00
G. W. Codwise	Confidential assistant	4,400 per year		$31 10	31 10
T. S. Bailey	Assistant engineer	2,580 per year	236 50		236 50
J. C. Bell	Assistant engineer	2,580 per year	409 89	58 55	468 44
G. E. Gibson	Assistant engineer	2,580 per year	666 50	82 44	748 94
T. R. Haselum	Assistant engineer	2,160 per year	363 00		363 00
H. W. Hale	Assistant engineer	2,580 per year	12 58	4 16	16 74
E. C. Olcott	Assistant engineer	2,208 per year	12 88	0 50	13 38
C. R. Waters	Assistant engineer	2,340 per year	208 45		208 45
C. T. Kniskern	Junior assistant engineer	1,800 per year	134 87		134 87
Samuel Levine	Junior assistant engineer	1,800 per year	40 14		40 14
L. E. Turpit	Junior assistant engineer	1,560 per year	105 69		105 69
Louis Cicio	Boatman	3 00 per day	3 30		3 30
H. J. Richardson	Photographer	1,680 per year		19 14	19 14
Post Master, Albany, N. Y.				200 00	200 00
Total			$2,314 80	$395 89	$2,710 69

Survey of Lands Under Water
Chapter 12, Laws of 1918

NAME	Rank	Rate of compensation	* Services	Expenses	Total
Edward Anderberg	Senior assistant engineer	$3,540 per year	$20 89		$20 89
Horace Corbin	Assistant engineer	2,580 per year	15 77		15 77
H. W. Hale	Assistant engineer	2,580 per year	115 61	$73 85	189 46
Edwin Hilborn	Assistant engineer	2,580 per year	139 53	3 28	142 81
L. T. Howard	Assistant engineer	2,580 per year	15 77		15 77
H. T. Arnold	Junior assistant engineer	1,800 per year	11 00		11 00
H. T. Eagan	Junior assistant engineer	1,800 per year	9 93		9 93
John Edelstein	Junior assistant engineer	1,680 per year	9 39		9 39
F. E. Gillen	Junior assistant engineer	1,800 per year	85 34		85 34
B. A. Krotinger	Junior assistant engineer	1,680 per year	4 52		4 52
Samuel Levine	Junior assistant engineer	1,800 per year	221 94	76 25	298 19
Charles Montag	Junior assistant engineer	1,800 per year	15 40		15 40
Isie Spahn	Junior assistant engineer	1,320 per year	18 51		18 51
Isaac Stern	Junior assistant engineer	1,440 per year	35 65		35 65
Joseph Hoehlein	Engineering assistant	840 per year	24 83		24 83
J. J. Raup	Engineering assistant	1,080 per year	9 58		9 58
James Daly	Laborer	2 50 per day	5 50		5 50
Total			$759 16	$153 38	$912 54

* Includes additional compensation of 10 per cent allowed above base rate.

Jamaica Bay–Peconic Bay Canal
Chapter 317, Laws of 1917; chapter 343, Laws of 1918

NAME	Rank	Rate of compensation	*Services	Expenses	Total
F. M. Williams	State Engineer and Surveyor	$8,000 per year		$90 35	$90 35
W. W. Wotherspoon	Superintendent of Public Works			27 45	27 45
Edward Anderberg	Senior assistant engineer	3,540 per year	$409 35	41 78	451 13
R. G. Finch	Chief clerk	4,200 per year		28 75	28 75
J. E. Stewart	Clerk	1,980 per year	165 00		165 00
Emily P. Hoffman	Stenographer	900 per year	82 50		82 50
H. E. Brainard	Assistant engineer	2,580 per year	236 50		236 50
P. H. Budd	Assistant engineer	2,208 per year	13 49	4 50	17 99
J. O. Burt	Assistant engineer	2,160 per year	552 75	922 30	1,475 05
Horace Corbin	Assistant engineer	2,580 per year	76 29		76 29
H. W. Hale	Assistant engineer	2,580 per year	141 44	31 99	173 43
W. S. McDowell	Assistant engineer	2,580 per year	236 50	21 60	258 10
Samuel Levine	Junior assistant engineer	1,800 per year	94 73		94 73
Charles Montag	Junior assistant engineer	1,800 per year	20 53		20 53
Isie Spahn	Junior assistant engineer	1,320 per year	150 98		150 98
G. W. Nostrand	Engineering assistant	900 per year	103 49		103 49
T. J. Torpy, Jr.	Engineering assistant	900 per year	182 42		182 42
Louis Cicio	Boatman	3 00 per day	13 20		13 20
W. B. Rowland	Boatman	3 00 per day	69 30		69 30
H. A. Dayton	Laborer	2 50 per day	176 00		176 00
Robert Harmon	Laborer	2 50 per day	231 00		231 00
J. R. Van Schoonhoven	Laborer	2 50 per day		17 95	17 95
H. J Richardson	Photographer	1,680 per year		32 75	32 75
Joseph Bailey				3 10	3 10
Total			$2,955 47	$1,222 52	$4,177 99

Mill River Survey
Chapter 427, Laws of 1918

NAME	Rank	Rate of compensation	*Services	Expenses	Total
Edward Anderberg	Senior assistant engineer	$3,540 per year	$11 34		$11 34
H. W. Hale	Assistant engineer	2,580 per year	93 66	$67 44	161 10
H. T. Arnold	Junior assistant engineer	1,800 per year	192 18		192 18
John Edelstein	Junior assistant engineer	1,680 per year	36 90		36 90
Isie Spahn	Junior assistant engineer	1,320 per year	25 98		25 98
Geo. W. Nostrand	Engineering assistant	900 per year	2 75		2 75
Louis Cicio	Boatman	3 00 per day	16 50		16 50
William Rowland	Boatman	3 00 per day	19 80		19 80
Total			$399 11	$67 44	$466 55

* Includes additional compensation of 10 per cent allowed above base rate.

Hydrographic Survey

Chapter 181, Laws of 1917; chapter 151, Laws of 1918
In coöperation with United States Geological Survey

William Alexander	$15 00
Alling Company	38 15
John Bisland	27 00
E. D. Burchard	256 52
Max H. Carson	202 36
W. E. Coe	15 00
H. B. Couch	45 00
C. C. Covert	381 52
C. S. De Golyer	51 00
W. & L. E. Gurley	567 75
E. W. Hart	15 00
O. W. Hartwell	583 71
Erastus Ingraham	45 00
Helen R. Kimmey	106 46
Leopold Voelpel & Co.	6 75
James Lyons	27 00
G. W. Marvin	45 00
Thos. M. Mills	48 00
J. Wendell Moulton	195 81
New York Telephone Company	29 90
D. L. Orcutt	36 00
C. L. Schenck	45 00
W. J. Shanly	36 00
Mrs. J. E. Sherman	24 00
Total	$2,932 93

SUMMARY

The foregoing tables are summarized as follows:

Ordinary Repairs to Canals

1. Erie canal, chapter 151, Laws of 1918	$6,577 24
2. Champlain canal, chapter 151, Laws of 1918	3,422 76

Construction of Barge Canal

3. Head office account, chapter 147, Laws of 1903, and amendatory laws	212,966 79
4. Erie canal, chapter 147, Laws of 1903, and amendatory laws	52,236 90
5. Champlain canal, chapter 147, Laws of 1903, and amendatory laws	34,015 76

Construction of Barge Canal Terminals

6. Barge canal terminals, chapter 746, Laws of 1911, and amendatory laws	155,691 78

Hudson River Terminals

7. Hudson river terminals, chapter 555, Laws of 1918	4,816 58

Bridge Designers, Engineers, etc.

8. Bridge designers, engineers, etc., chapter 151, Laws of 1918	2,000 00

Special Work

9. High street bridge, Cohoes, chapter 181, Laws of 1917; chapter 151, Laws of 1918	820 96
10. Schenectady–Scotia bridge, chapter 735, Laws of 1917; chapter 634 of 1919	3,356 02
11. Sea-wall, Orient-East Marion, L. I., chapter 428, Laws of 1918	739 52

Special Surveys

12. Blue line surveys, chapter 151, Laws of 1918	15,000 00
13. Surveys for State Court of Claims, chapter 151, Laws of 1918	200 00
14. Department surveys, chapter 151, Laws of 1918	3,627 93
15. State boundary, chapter 151, Laws of 1918	378 26
16. Delaware–Schoharie county boundary line, chapter 559, Laws of 1918	2,375 55
17. Saratoga–Warren county boundary line, chapter 561, Laws of 1918	2,476 80
18. Ulster–Greene county boundary line, chapter 562, Laws of 1918; chapter 600, Laws of 1919	3,682 69
19. Land grants, chapter 151, Laws of 1918	2,710 69
20. Survey of lands under water, chapter 12, Laws of 1918	912 54
21. Jamaica Bay–Peconic Bay canal, chapter 317, Laws of 1917; chapter 343, Laws of 1918	4,177 99
22. Mill river survey, chapter 427, Laws of 1918	466 55
23. Hydrographic survey, chapter 181, Laws of 1917; chapter 151 Laws of 1918	2,932 93
Total	$515,586 26

REPORT

OF THE

DIVISION ENGINEER

OF THE

MIDDLE DIVISION

For the Fiscal Year Ended June 30, 1919

[117]

MIDDLE DIVISION

STATE OF NEW YORK

DEPARTMENT OF STATE ENGINEER AND SURVEYOR

MIDDLE DIVISION

SYRACUSE, N. Y., *July* 1, 1919.

Hon. FRANK M. WILLIAMS, *State Engineer and Surveyor,*
Albany, N. Y.:

Sir.— I have the honor of submitting herewith my annual
report as Division Engineer of the Middle Division of the New
York State canals for the fiscal year ended June 30, 1919.

Owing to the practical completion of the Barge canal on this
Division, the engineering force has been much smaller than for
a number of preceding years. It has been employed in looking
after a few unfinished contracts and also some special appropriation work, in making surveys and maps for the appropriation and
release of lands affected by Barge canal construction in collect-
ing and preparing data for the Court of Claims and in locating
and mapping the blue lines of the old Erie canal, preparatory
to its abandonment, as detailed below.

In general the Barge canal channel has been of full width and
depth throughout the division, and far more free from bars and
obstructions than would naturally be expected. The amount of
material that was washed into the channel between lock No. 22
and Oneida lake was less than usual and was removed by the
Department of Public Works in such a manner as not to interfere
with navigation.

The canal structures on the Division are generally in good condition. Some repairs, however, are needed. In my report for
last year I called attention to leakage at the Curved dam at
Oswego and at the dam across the Owasco outlet at Auburn, stat-
ing that repairs should be made.

While there has been no trouble the past year in maintaining the level of the old Erie canal between Syracuse and New London, owing to the wet season, a dry season would tax the reservoirs and feeders severely. It is of the utmost importance that the State be in a position to conserve the waters of the canal system.

The dam across Chittenango creek, at the head of the feeder, is in exceedingly bad condition. It is a wooden structure with masonry abutments and bulkhead walls. It leaks badly, both through and underneath the spillway. It is liable to go out at any time. It should be replaced with a new concrete structure at the earliest possible date.

Cowasselon dam was washed out several years ago. It should be rebuilt in concrete, to divert the water from creek to the feeder.

During the past season the Madison reservoirs have not been used to any extent, owing to the condition of the feeders, which have become so filled with vegetation and slit as to make them almost useless. These reservoirs supply the level of the old canal between Rome and Utica. Although a culvert is being placed under the Barge canal at Rome, to supply this level from the Delta reservoir *via* the Mohawk river, nevertheless, all possible use should be made of the Madison reservoirs. But to do so, the feeders should be cleaned out. This is earnestly recommended.

Each year a large volume of sand is washed into the Barge canal channel by Wood creek and other streams between lock No. 22 and Sylvan Beach, making almost constant dredging necessary. As a partial remedy it is suggested that the banks of the streams entering the canal in this section be planted with willows. As soon as thoroughly rooted the willows will prevent further erosion of the stream banks and consequent deposits of sand in the canal channel.

It seems to be almost impossible to control the speed of the lighter sort of boats using the canal. This is especially true on long reaches between structures. As a consequence in land sections, especially where the prism is in embankment, serious damage has been done. In places where they are formed of light material the banks have been eroded nearly half their thickness at the water-line. It is a self-evident fact that sooner or later all such banks will have to be protected with stone. For this reason it is recommended that all stone or rock excavated from the canal and placed in spoil-banks adjacent thereto be carefully conserved, as I am satisfied that every yard of such stone will be needed by the State for bank protection in the near future.

Attention is called to the concrete in some of the Barge canal structures. In some instances, notably at lock 24 and at Delta, disintegration of the concrete has become a serious matter. It has progressed to such an extent that extensive repairs are needed at once. The cause of the disintegration is not definitely known at this writing and a most thorough study and careful analysis of the conditions and materials used in construction is recommended, to the end that not only the cause but a remedy may be found.

The appended reports of the Senior Assistant Engineers will give in detail the progress and condition of the work on the Division. There is appended also a tabulation showing the name, rank and compensation of the men on the Division, together with the incidental expenses for the fiscal year.

I heartily thank you and my other superior officers for your courtesy and assistance in performing the work on the Division, and commend the men under me for their support and faithful and efficient service.

Respectfully submitted,

GUY MOULTON,
Division Engineer.

APPENDED REPORTS — MIDDLE DIVISION

SPECIAL APPROPRIATIONS

Improvement of Cowasselon Creek, in the County of Madison, by Dredging and Otherwise

(Chapter 781, Laws of 1917)

Contractor, Robert Provo.
Engineer in charge, David R. Lee.

Engineer's estimate	$12,000.00
Contractor's bid	10,500.00
Amount of final account	9,572.70

Construction of a New Plate Girder Bridge over the Black River Canal at Whitesboro Street, Rome

(Chapter 753, Laws of 1917)

Contractor, Walter S. Rae.
Engineer, William J. Durken.

Engineer's estimate	$12,085.00
Contractor's bid	11,683.00
Work done to date.........................	10,840.00

Construction of a Dive Culvert at Rome

(Chapter 346, Laws of 1918)

Contractors, Scott Bros.
Engineer in charge, Foster B. Crocker.

Engineer's estimate	$42,811.20
Contractor's bid	46,731.20

This contract is for the construction of an 8-foot pipe culvert under the Barge canal at Rome, to furnish water to the section of the Erie canal between Rome and Utica.

Contract work has not been started.

Land Abandonments and Surveys

(Chapter 299, Laws of 1916)

Surveys have been made and maps prepared for the abandonment of the old Erie and Oswego canals and all State lands adjacent thereto, all according to rules and regulations laid down by the Commissioners of the Land Office, at the following locations:

Erie canal, between Third and Schuyler streets in the city of Utica.

Erie canal, between Centerport and the Wayne county line.

Oswego canal, from the junction of the Erie and Oswego canals to the north city line of the city of Syracuse.

Oswego canal, lands occupied by the North side-cut.

Oswego canal, lands occupied by the South side-cut.

Court of Claims

In addition to the work usually required by the Superintendent of Public Works in connection with ordinary repairs, a large amount of survey work and mapping has been made of property alleged to have been damaged by the State, reports made and data properly arranged for the Court of Claims and the Attorney-General. A large amount of time is spent by the engineers as witnesses for the State on local claims during the sessions of the Court.

Blue Line Surveys

(Chapter 646, Laws of 1916; chapter 151, Laws of 1918)

These laws provide for surveys, field notes and manuscript maps affecting various canals and canal lands.

Two field parties were constantly at work establishing and monumenting the State's right of way along the Erie and Oswego canals.

One field party, engaged in construction work, at such times as supervision of construction would permit worked intermittently on blue line surveys, establishing the State's right of way in the vicinity of Phoenix.

Field work started July 8, 1918, and has continued since.

The following work has been done during the fiscal year ended June 30, 1919:

Erie Canal

The blue and red lines were established and monumented from the Oneida–Herkimer county line to the west city line of Utica and tracings on standard-sized sheets were completed. These maps were approved by the Canal Board, May 21, 1919.

The red and blue lines were established and monumented from the west corporation line of Rome to New London and tracings on standard-sized sheets made. These maps were approved by the Canal Board, August 6, 1919.

The red and blue lines have been established through the city of Rome. No monuments have as yet been set. The map, on a scale of 1 inch = 100 feet, is about 90 per cent. completed.

The following is a summary of work done between the west city line of Utica and the east corporation line of the city of Rome, a distance of 11.9± miles, during the fiscal year:

Base line run and topography taken over entire line.

Red line established from the west city line of Utica to Clinton street bridge, Whitesboro, a distance of 1½ miles.

Map, on a scale of 1 inch = 100 feet, about 30 per cent platted.

Oswego Canal

The red and blue lines have been established and monumented through the.village of Phoenix. Tracings on standard-sized sheets have been completed.

ERIE CANAL, RESIDENCY NO. 5.

Senior Assistant Engineer Edward J. Berry reports:

This residency extends from the east end of Oneida county to Oneida lake, a distance of 31.06 miles, and includes the former water-supply residency.

In connection with Barge canal construction, our engineers have been actively engaged in assisting the department of the Superintendent of Public Works, making surveys, sweeping channel, setting buoys and general maintenance work.

Reports are given on contracts Nos. 42-A, 44-A, 156 and 187, terminal contracts Nos. 15-M, 63, and 220, and parts of terminal contracts Nos. 101, 106 and 109.

Contract No. 42-A

This contract is for completing the construction of the canal, together with all incidental work, between the Herkimer–Oneida county line and a point just east of Oriskany road, Sta. 5775. Length, 8.96 miles. It was awarded to Grant Smith & Co. & Locher, being signed on February 24, 1913. Construction work began March 13, 1913. The engineer's preliminary estimate was $1,033,037.85, the contractor's bid, $1,014,671.83. The contract price as modified by alterations Nos. 1 and 2 is $1,239,045.03. Excess steel castings to the value of $220.00 have been authorized by the Canal Board. The work was accepted July 24, 1918, and the final account, amounting to $1,197,244.78, was approved by the Canal Board March 5, 1919. The amount paid on extra work orders to date is $332.69.

Contract work was completed prior to June 30, 1918.

Contract No. 44-A

This contract is for completing the construction of the canal prism near the junction lock at New London. It was awarded to Scott Brothers, being signed on October 10, 1916. Construction work began about November 1, 1916. The engineer's preliminary estimate was $57,050.00, the contractor's bid, $52,486.00.

Foster B. Crocker, Assistant Engineer, was in charge.

Under authority of chapter 585, Laws of 1918, this contract was canceled by the Canal Board on July 9, 1918. The cancelation became effective August 31, 1918, upon approval of the Canal Board, the contractor having filed a stipulation of his compliance with the terms of the law. The actual cost of the work from April 7, 1917, to August 31, 1918, was $40,946.71, and payment of balance due on this amount was authorized by the Canal Board on March 19, 1919. The final account for work done prior to April 7, 1917, amounting to $15,295.46, was approved by the Canal Board on March 19, 1919.

This contract has been completed and the total payment, including extra work orders, was $56,242.17.

After a discontinuance of about six months, work was resumed on this contract on July 29, 1918. At that time the State dipper-dredge *Pathfinder* started to work removing material above grade

in prism. This machine worked until August 24, 1918, removing
over 12,000 cubic yards of material, which was towed to Oneida
lake and spoiled in deep water. This completed the contract.

Contract No. 187

This contract is for placing wash-wall protection between New
London and lock No. 22. It was awarded to Scott Brothers, being
signed on August 20, 1918. The engineer's preliminary estimate
was $17,525.00, the contractor's bid, $22,530.00. The value of
work done during the year is $15,470.00.

F. B. Crocker, Assistant Engineer, is in charge.

The placing of wash-wall protection under this contract has been
completed, but the work is still to be accepted and final account
rendered.

Contract No. 156

This contract is for constructing a highway bridge across Wood
creek about one mile east of Sylvan Beach. It was awarded to
Chesley, Earl & Heimbach, Inc., being signed on August 28, 1917.
Construction work began October 17, 1917. The engineer's pre-
liminary estimate was $7,788.00, the contractor's bid, $9,813.00.
The contract price as modified by alteration No. 1, is $10,113.00.
Excess quantities to the value of $431.50 have been authorized by
the Canal Board. The value of work done during the year is
$7,133.38. The work was accepted December 27, 1918, and the
final account, amounting to $9,643.30, was approved by the Canal
Board February 13, 1919.

F. B. Crocker, Assistant Engineer, was in charge.

During the fiscal year the concrete abutments were poured, steel
superstructure erected, bridge approaches finished and the bridge
thrown open to traffic.

Terminal Contract No. 15-M — Utica

This contract is for electrical equipment and machinery for
operating and lighting the Utica terminal lock. It was awarded to
Lupfer & Remick, being signed on October 31, 1917. Construction
work began April 26, 1918. The engineer's preliminary estimate
was $30,681.20, the contractor's bid, $36,967.50. Excess quantities to the value of $123.12 have been authorized by the Canal

Board. The value of work done during the year is $29,929.72. The work was accepted May 21, 1919, and the final account, amounting to $37,069.72, was approved by the Canal Board June 25, 1919.

Lewis Bartlett, Assistant Engineer, was in charge.

During the fiscal year the contractors have completed the erection and testing of all the valve and gate machinery and have placed the electric lights on the lock.

Terminal Contract No. 63 — Utica

This contract is for constructing railroad tracks and brick pavement at the Utica terminal. It was awarded to Harry W. Roberts & Co., being signed on April 19, 1918. The engineer's preliminary estimate was $9,590.00, the contractor's bid, $10,164.00. The contract price as modified by alteration No. 1 is $7,672.00. Excess quantities to the value of $277.00 have been authorized by the Canal Board. The value of work done during the year is $7,632.13. The work was accepted November 13, 1918, and the final account, amounting to $7,632.13, was approved by the Canal Board November 13, 1918. The amount paid on extra work orders during the year is $632.17, total to date, the same.

Lewis Bartlett, Assistant Engineer, was in charge.

Alteration No. 1, approved by the Canal Board July 9, 1918, provides for eliminating railroad tracks and bumping posts and for changing location of pavement. It decreases the contract price by $2,492.00.

An extra work order dated May 8, 1918, provides for making certain alterations to the freight-shed, so as to provide a field office for the engineers at the site. The final account, amounting to $414.63, was approved by the Canal Board August 14, 1918.

An extra work order dated August 12, 1918, provides for building a catch-basin in order to get proper drainage along the roadway leading to the terminal, and also for removing the electric service line pole which stood in the center of the proposed pavement. The final account, amounting to $217.54, was approved by the Canal Board August 31, 1918.

An extra work order dated October 17, 1918, provides for extending the pavement, in order to provide access to the extension to the warehouse built under terminal contract No. 220.

During the fiscal year the catch-basin and tile drain were completed, the concrete sidewalk abutting North Genesee street was laid and the brick pavement finished.

Terminal Contract No. 220 — Utica

This contract is for constructing an extension to the Utica terminal freight-house. It was awarded to James T. Young, being signed on August 12, 1918. Construction work began August 14, 1918. The engineer's preliminary estimate was $5,000.00, the contractor's bid, $5,495.00. The value of work done during the year is $5,324.40. The work was accepted October 16, 1918, and the final account, amounting to $5,324.40, was approved by the Canal Board November 13, 1918.

Lewis Bartlett, Assistant Engineer, was in charge.

This contract added an extension 32 feet by 100 feet to the existing freight-house. It was completed by September 25, 1918.

Terminal Contract No. 106

This contract is for furnishing fourteen two-ton steam tractor cranes for Barge canal terminals. The following report relates to the work at Utica. The contract was awarded to the John F. Byers Machine Co., being signed on February 14, 1918. The engineer's preliminary estimate was $5,250.00 per crane, the contractor's bid, $5,265.00 per crane. The contract price as modified by alteration No. 1 is $5,515.00 per crane. The work was accepted September 24, 1918, and the final account, amounting to $5,515.00 for Utica, was approved by the Canal Board October 9, 1918.

Lewis Bartlett, Assistant Engineer, was in charge.

The contractors delivered the maintenance accessories during the month of July, 1918, which completed the contract.

Terminal Contract No. 109

This contract is for furnishing electric capstans and trolley hoists at Pier 6, East river, and West 53d street pier, New York city, and electric capstans at the Utica terminal lock. The following report relates to the work at Utica. The contract was awarded to the General Electric Co., being signed August 2, 1918. The

CONTROLLING WORKS IN THE OSWEGO RIVER AT FULTON (UPPER DAM)

Beside the fixed dam there are Taintor gates and on each side head-gates for an industrial power-plant. The view shows all the Taintor gates open for passing a spring flood.

DAM ACROSS THE SENECA RIVER AT BALDWINSVILLE

View during a spring flood. At the right end of the dam is seen an automatic flood gate.

DAM ACROSS THE OSWEGO RIVER AT MINETTO

View during a spring flood, when 20,000 cubic feet of water per second were flowing.

DAM ACROSS THE OSWEGO RIVER SOUTH OF OSWEGO
Known as new High dam. View during a spring flood.

CONTROLLING WORKS AT FOOT OF CAYUGA LAKE

By means of these Taintor gates the flow of Seneca river and the water in Cayuga lake are controlled.

CANAL STRUCTURES AT WATERLOO

At the right is the lock with a guard-gate at its head. At the left are Taintor gates, which constitute controlling works to regulate the water in Seneca river and also in Seneca lake.

CANAL STRUCTURES AT WATERLOO

In the foreground is a forebay, which carries water to various industrial power-plants. The bridge crosses the forebay and also Seneca river, at the right. Beyond are the Taintor gate controlling works and the guard-gate at the head of the lock.

LAKE STREET BRIDGE, GENEVA

This is a plate girder bridge. The view is over the bridge, looking up Lake street.

LAKE STREET BRIDGE, GENEVA

Side view. This bridge spans the old Cayuga and Seneca canal and was constructed under authority of chapter 351, Laws of 1918.

engineer's preliminary estimate was $1,500.00 per capstan, the contractor's bid, $1,463.00 per capstan.

Lewis L. Bartlett, Assistant Engineer, is in charge.

No capstans have been delivered at Utica.

Terminal Contract No. 101

This contract is for furnishing and installing steel stiff-leg derricks at Albany, Whitehall, Little Falls, Rome, Lockport and Tonawanda. The following report relates to work at Rome. The contract was awarded to E. Brown Baker, being signed on December 18, 1916. On February 21, 1917, it was assigned to the Mohawk Dredge & Dock Co., Inc., and this assignment was approved by the Superintendent of Public Works March 26, 1917. The engineer's preliminary estimate for derrick at Rome was $3,885.50, the contractor's bid, $5,684.00. Excess metal (Rome) to the value of $1,330.00 has been authorized by the Canal Board. The work was accepted December 4, 1918, and the final account, amounting to $6,831.48, was approved by the Canal Board December 27, 1918.

L. W. Bartlett, Assistant Engineer, was in charge.

The work at Rome had been completed a year ago, but the acceptance and approval of the final account occurred during the past fiscal year.

ERIE CANAL, RESIDENCIES NOS. 6 AND 7

Senior Assistant Engineer Edward J. Berry reports:

Residency No. 6 extends from deep water at the western end of Oneida lake to Baldwinsville, a distance of 23.4 miles, and includes also the work under contract No. 132, which pertains to aids to navigation on Oneida lake.

Residency No. 7 extends from Baldwinsville to the Wayne county line, a distance of 32.7 miles.

Reports are given on the following contracts: Contracts Nos. 46-B, 165, 184 and 188, and a part of No. 172, and terminal contracts Nos. 20, 28 and 213, and a part of No. 106.

Contract No. 184

This contract is for excavating a channel under the N. Y. C. R. R. bridge at Brewerton. It was awarded to Mohawk Dredge & Dock Co., Inc., being signed on April 12, 1918. The engineer's preliminary estimate was $7,200.00, the contractor's bid, $9,480.00. The work was accepted July 9, 1918, and the final account, amounting to $9,562.95, was approved by the Canal Board August 14, 1918.

Contract work was finished on June 21, 1918, prior to the report of a year ago, but the acceptance and approval of the final account occurred during the past fiscal year.

Contract No. 188

This contract is for completing the canal prism excavation at the N. Y. C. R. R. bridge, Brewerton. It was awarded to E. Brown Baker, being signed on August 7, 1918. The engineer's preliminary estimate was $30,000.00, the contractor's bid, $35,400,00. The value of work done during the year is $30,260.00, total done to date, the same.

H. H. Brown, Assistant Engineer, is in charge.

The work has been completed except for a small amount of material that will require drilling and blasting.

Contract No. 165

This contract is for removing the Montezuma aqueduct and completing the canal prism excavation from Sta. 5439 + 48, just east of the aqueduct, to Sta. 5550, near May's Point, and for redredging the canal prism near Fox Ridge. It was awarded to Mohawk Dredge & Dock Co., Inc., being signed on November 23, 1917. Construction work began in December, 1917. The engineer's preliminary estimate was $84,530.00, the contractor's bid, $160,943.00. The value of work done during the year is $29,118.26. The work was accepted January 15, 1919, and the final account, amounting to $145,798.26, was approved by the Canal Board April 2, 1919. The amount paid on extra work orders during the year is $315.69, total to date, the same.

J. G. Palmer, Assistant Engineer, was in charge.

An extra work order dated February 11, 1918, provides for peeling, painting and bolting together fender piles driven at railroad bridges. The final account, amounting to $315.69, was approved by the Canal Board September 24, 1918.

During the fiscal year the prism excavation was completed, all aqueduct masonry, wooden trunk and foundation timbers removed and all foundation piles pulled, thus completing the contract.

Contract No. 46-B

This contract is for completing the construction of a lock, dam, etc., at May's Point. Length, 0.66 mile. It was awarded to Scott Bros., being signed on February 25, 1916. The engineer's preliminary estimate was $314,660.72, the contractor's bid, $277,-348.22. The contract price as modified by alterations Nos. 1 and 2 is $293,676.97. Excess quantities to the value of $3,927.71 have been authorized by the Canal Board. The value of work done during the year is $10,638. The work was accepted April 16, 1919, and the final account, amounting to $269,398.41, was approved by the Canal Board July 16, 1919. The amount paid on extra work orders during the year is $1,451.08, total to date, $1,575.63.

J. G. Palmer, Assistant Engineer, was in charge.

An extra work order dated July 24, 1918, provides for a change in plan of building snubbing-posts by drilling holes for and furnishing and placing dowels, building forms for concrete settings and removing a portion of the backfill already placed; also for payment for piling rendered useless by the changed plans. The final account, amounting to $260.65, was approved by the Canal Board December 10, 1918.

An extra work order dated December 26, 1918, provides for moving derrick, laying track, housing needles, lumber delivered and closing openings in steel sheet-piling. The final account, amounting to $976.58, was approved by the Canal Board February 13, 1919.

An extra work order dated March 20, 1919, provides for cutting away parts of Z-bars which rest upon wicket stiffener angles and attaching oak planking to lower gates of the movable dam at May's

Point. The final account, amounting to $213.85, was approved by the Canal Board July 25, 1919.

During the fiscal year the embankment at the south approach to the movable dam was placed, chains and counterweights for lower gates and wickets instaled and embankments covered with top soil and seeded.

Contract No. 172

This contract is for furnishing and delivering barrel buoys and lamp posts for aid to navigation on the Seneca, Clyde, Genesee and Tonawanda rivers. It was awarded to Lupfer & Remick, being signed on March 15, 1918. The engineer's preliminary estimate for the whole contract was $14,853.00, the contractor's bid, $13,063.20. The contract price as modified by alteration No. 1 is $12,921.45. The value of work done during the year is $1,693.35. The work was accepted September 24, 1918, and the final account, amounting to $12,913.35, was approved by the Canal Board September 24, 1918. The amount paid on extra work orders to date is $906.50, of which amount $465.50 applies to work on this residency.

N. R. McLoud, Junior Assistant Engineer, was in charge.

The final estimate for work on this residency amounted to $6,398.45.

Work on this residency was completed during the summer of 1918.

Terminal Contract No. 28 — Cleveland

This contract is for constructing a harbor, dockwall and two breakwaters on Oneida lake at Cleveland. It was awarded to Clarence E. Gruner, being signed on February 15, 1915. It was assigned to Barrally & Ingersoll and this assignment was approved by the Superintendent of Public Works March 15, 1915. Construction work began in June, 1915. The engineer's preliminary estimate was $34,575.00, the contractor's bid, $30,673.00. The contract price as modified by alterations Nos. 1 and 2 is $37,-222.00. Excess quantities to the value of $1,675.00 have been authorized by the Canal Board. The value of work done during the year is $3,010.00, total done to date, $35,120.

W. J. Durkan, Assistant Engineer, is in charge.

During the fiscal year seven concrete tops were placed on the east breakwater pier, finishing that pier. With the exception of a few small areas above grade in the channel contract work is completed.

Terminal Contract No. 20 — Syracuse

This contract is for constructing a terminal basin with a connecting channel to Onondaga lake, also piers, dockwalls, spillway and a highway bridge at Syracuse. It was awarded to the Walsh Construction Co., Inc., being signed on November 4, 1915. Construction work began in same month. The engineer's preliminary estimate was $665,875.00, the contractor's bid, $419,-659.00. The contract price as modified by alterations Nos. 1, 2, 3 and 4 is $549,878.26. Excess sheeting and bracing to the value of $25,200.00 has been authorized by the Canal Board. The amount paid on extra work orders to date is $1,174.50.

A. G. Card, Assistant Engineer, is in charge.

Alteration No. 4, approved by the Canal Board December 27, 1918, provides for eliminating riprap along N. Y. C. R. R. embankment facing the lake front. It decreases the contract price by $16,875.00.

Under authority of chapter 585, Laws of 1918, this contract was canceled by the Canal Board on December 27, 1918. The cancelation became effective February 13, 1919, on approval of the Canal Board, the contractor having filed a stipulation of his compliance with the terms of the law. The actual cost of the work from April 7, 1917, to February 13, 1919, was $285,727.10, and payment of balance due on this amount was authorized by the Canal Board on May 21, 1919. The final account for work done prior to April 7, 1917, amounting to $358,981.95, was approved by the Canal Board on May 7, 1919.

This contract has been completed and the total payment, including extra work orders, was $645,484.

During the fiscal year the clearing and removing of buildings on terminal site was completed, channel excavation completed north of Hiawatha street, riprap laid at the north end of the west abutment of the N. Y. C. R. R. bridge and six clusters

of guide piles driven in Onondaga lake. Excavation beyond the
limits of surfacing was completed. All cinder surfacing on piers
and along walls and approach was laid 24 inches thick and rolled.
Fender piles were driven on corners of piers and fender timbers
placed on dockwalls for the entire length.

Terminal Contract No. 213 — Syracuse

This contract is for constructing a frame freight-house and
four electrically-operated timber derricks at Syracuse. It was
awarded to the Savage Construction Co., being signed on February
14, 1918. Construction work began in February, 1918. The
engineer's preliminary estimate was $28,200.00, the contractor's
bid, $27,032.00. The contract price as modified by alteration
No. 1 is $26,997.00. The value of work done during the year is
$11,816.40. The work was accepted May 21, 1919, and the final
account, amounting to $26,346.40, was approved by the Canal
Board June 25, 1919. The amount paid on extra work orders
during the year is $371.82, total to date, the same.

A. G. Card, Assistant Engineer, is in charge.

An extra work order dated January 3, 1919, provides for installing transmission line from service line of Syracuse Lighting Co. to
the freight-house and furnishing and installing transformers.
The final account, amounting to $371.82, was approved by the
Canal Board June 11, 1919.

During the fiscal year the carpenter work and painting were
finished, the derrick hoists, motors and controllers delivered and
placed, and the transmission line erected and tested. The freight-
house was opened for use in September.

Terminal Contract No. 106

This contract is for furnishing fourteen two-ton steam tractor
cranes for Barge canal terminals. One of these is for the Syracuse
terminal. The contract was awarded to the John F. Byers
Machine Co., being signed on February 14, 1918. The engineer's preliminary estimate was $5,250.00 per crane, the con-
tractor's bid, $5,265.00 per crane. The contract price as modi-
fied by alteration No. 1 is $5,515.00 per crane. The work was

accepted September 24, 1918, and the final account, amounting to $5,515.00 for Syracuse, was approved by the Canal Board October 9, 1918.

A. G. Card, Assistant Engineer, was in charge.

The crane for the Syracuse terminal was delivered during the past fiscal year.

OSWEGO CANAL RESIDENCY

Senior Assistant Engineer Edward J. Berry reports:

This residency comprises all work on the Oswego canal. Reports are given on contracts Nos. 99, 117, 167 and 182, and terminal contracts Nos. 30, 33-P, 59, 60 and 226, and part of No. 106.

Contract No. 167

This contract is for constructing a bascule bridge below lock No. 1, at Culvert street, Phoenix. It was awarded to Walter S. Rae, being signed on October 13, 1917. Construction work began in May, 1918. The engineer's preliminary estimate was $26,653.60, the contractor's bid, $29,689.30. The value of work done during the year is $17,460.00, total done to date, $18,460.00.

N. R. McLoud, Junior Assistant Engineer, is in charge.

All the excavation has been completed, the abutments and approaches built, steel superstructure erected and operating machinery partially installed.

Contract No. 117

This contract is for constructing a swing-bridge over lock No. 2 at Fulton. It was awarded to Walter S. Rae, being signed on April 15, 1918. Construction work began October 21, 1918. The engineer's preliminary estimate was $34,713.30, the contractor's bid, $36,513.80. The value of work done during the year is $11,570, total done to date, the same.

H. H. Brown, Assistant Engineer, is in charge.

The main pivot foundation and bridge lock have been completed. Concrete work in the east and west abutments is well under way.

Contract No. 99

This contract is for constructing portions of a bridge over the Oswego river at Minetto. It was awarded to Larkin & Sangster, being signed on September 12, 1916. The engineer's preliminary estimate was $117,170.75, the contractor's bid, $115,980.75.

Edward M. Ellis, Assistant Engineer, was in charge.

Under authority of chapter 585, Laws of 1918, this contract was canceled by the Canal Board on July 9, 1918. The cancelation became effective December 10, 1918, upon approval of the Canal Board, the contractor having filed a stipulation of his compliance with the terms of the law. The actual cost of the work from April 7, 1917, to December 10, 1918, was $141,096.59, and payment of balance due on this amount was authorized by the Canal Board on February 13, 1919. The final account for work done prior to April 7, 1917, amounting to $11,250.65, was approved by the Canal Board on February 13, 1919.

This contract has been completed and the total payment, including extra work orders, was $153,086.24.

During the past year the sidewalks were completed, the pavement laid, riveting of new steel superstructure finished and the site cleaned up.

Contract No. 182

This contract is for completing excavation in front of the terminal dockwall below lock No. 8, Oswego. It was awarded to E. Brown Baker, being signed on August 30, 1918. Construction work began November 21, 1918. The engineer's preliminary estimate was $28,215.00, the contractor's bid, $30,267.00. The value of work done during the year is $26,660.00.

George Haley, Assistant Engineer, is in charge.

Contract work is completed with the exception of the removal of a few high spots.

Terminal Contract No. 30 — Oswego, River Terminal

This contract is for constructing a dockwall, an approach to the terminal and appertaining structures on the east side of the Oswego river between Schuyler and Cayuga streets, Oswego. It was awarded to Henry P. Burgard, being signed on March 24, 1916. Construction work began in April, 1916. The engineer's preliminary estimate was $103,700.00, the contractor's bid, $90,-984.00. The contract price as modified by alterations Nos. 1, 2 and 3 is $106,166.70. Excess quantities to the value of $943.00 have been authorized by the Canal Board. The work was accepted June 26, 1918, and the final account, amounting to $100,382.70, was approved by the Canal Board September 24, 1918. The amount paid on extra work to date is $1,406.50.

George H. Haley, Assistant Engineer, was in charge.

Construction work had been completed and accepted prior to July 1, 1918, but the final account was approved during the past fiscal year.

Terminal Contract No. 106

This contract is for furnishing fourteen two-ton steam tractor cranes for Barge canal terminals. One of these is for use at Oswego. The contract was awarded to the John F. Byers Machine Co., being signed on February 14, 1918. The engineer's preliminary estimate was $5,250.00 per crane, the contractor's bid, $5,265.00 per crane. The contract price as modified by alteration No. 1 is $5,515.00 per crane. The value of the work done at Oswego during the year is $1,305.00. The work was accepted September 24, 1918, and the final account, amounting to $5,515.00 for the work at Oswego, was approved by the Canal Board, October 9, 1918.

The crane had been delivered at Oswego prior to a year ago, but the work was accepted and the final account approved during the past fiscal year.

Terminal Contract No. 226 — Oswego, River Terminal

This contract is for constructing a frame freight-house and compacting gravel surfacing on the river terminal, Oswego. It was awarded to J. A. Laporte, being signed on April 28, 1919. Construction work began May 13, 1919. The engineer's preliminary estimate was $6,000.00, the contractor's bid, $5,199.00. The value of work done during the year is $5,030.

Construction work was practically completed during the month of June, 1919.

Terminal Contract No. 59 — Oswego, Lake Terminal

This contract is for constructing a railroad track approach to the terminal pier at Oswego. It was awarded to W. F. Martens, being signed on May 6, 1918. Construction work began in May, 1918. The engineer's preliminary estimate was $5,100.00, the contractor's bid, $6,516.00. The value of work done during the year is $3,771.41. The work was accepted August 31, 1918, and the final account, amounting to $5,391.41, was approved by the Canal Board December 27, 1918.

George H. Haley, Assistant Engineer, was in charge.

During the year the placing of ballast was done, the tracks brought to final alignment and contract work completed.

Terminal Contract No. 60 — Oswego, Lake Terminal

This contract is for constructing railroad and crane tracks on the terminal pier at Oswego. It was awarded to W. F. Martens, being signed on May 6, 1918. Construction work began in May 1918. The engineer's preliminary estimate was $8,365.00, the contractor's bid, $9,690.00. The value of work done during the year is $5,709. The work was accepted August 31, 1918, and the final account, amounting to $9,119.00, was approved by the Canal Board December 27, 1918.

George H. Haley, Assistant Engineer, was in charge.

During the year the placing of ballast was done, the tracks brought to final alignment and contract work completed.

Terminal Contract No. 33-P — Oswego, Lake Terminal

This contract is for paving part of the terminal pier at Oswego. It was awarded to Guy B. Dickison, being signed on May 7, 1918. Construction work began August 19, 1918. The engineer's preliminary estimate was $11,010.00, the contractor's bid, $11,730.00. Excess excavation to the value of $240.00 has been authorized by the Canal Board. The value of work done during the year is $11,329.00. The work was accepted October 16, 1918, and the final account, amounting to $11,329.00, was approved by the Canal Board January 15, 1919.

George H. Haley, Assistant Engineer, was in charge.

Contract work was completed September 26, 1918.

Cayuga and Seneca Canal Residency

Senior Assistant Engineer H. C. Smith reports:

This residency comprises all the work on the Cayuga and Seneca canal. Reports follow on contracts M, Q, R, T and U. The construction of bridges at Lake street, Geneva, and of a concrete dockwall at Canandaigua lake harbor, authorized by special acts of the Legislature, have been under the supervision of this office. Reports on this work follow those on the Barge canal contracts.

Contract R

This contract is for completing the unfinished work at several locations on the Cayuga and Seneca canal. It was awarded to the Sherman-Stalter Company, being signed on April 30, 1918. Construction work began May 22, 1918. The engineer's preliminary estimate was $185,259.00, the contractor's bid, $180,122.-80. The value of work done during the year is $158,264.38. The work was accepted May 21, 1919, and the final account, amounting to $173,434.38, was approved by the Canal Board August 20, 1919.

L. L. Hadley, Assistant Engineer, was in charge.

During the year prism excavation was completed at Mud lock, at Cayuga, at the N. Y. C. R. R. bridge 2½ miles east of Seneca Falls and at the Lehigh Valley R. R. bridges near Seneca lake. Temporary highway bridges were removed at Lake road, at Kingdon bridge west of Seneca Falls and at Free bridge. Excavation for the approach wall at Mud lock was made, the wall built and backfill placed. Construction work was completed on April 29, 1919.

Contract M

This contract is for constructing power-plants and for furnishing and installing electrical equipment and machinery for operating and lighting locks Nos. 1, 2, 3 and 4. It was awarded to Lupfer & Remick, being signed on November 5, 1914. Construction work began January 23, 1915. The engineer's preliminary estimate was $176,087.00, the contractor's bid, $188,-031.00. The contract price as modified by alterations Nos. 1 and 2 is $191,405.00. Excess chipping concrete to the value of $410.40 has been authorized by the Canal Board. The work was accepted December 4, 1918, and the final account, amounting to $190,274.64, was approved by the Canal Board December 4, 1918. The amount paid on extra work orders during the year is $221.35, total to date, the same.

Alteration No. 2, approved by the Canal Board September 24, 1918, provides for eliminating certain work from the contract. It decreases the contract price by $31.00.

The final account of extra work order dated March 16, 1918, amounting to $221.35, was approved by Canal Board December 27, 1918. The only work done during the year was in completing the changing of racks under this work order.

Contract T

This contract is for extending the core wall and other work at north end of dam No. 2, Seneca Falls. It was awarded to Kennedy & Scullen Construction Co., Inc., being signed on January 20, 1919. Construction work began March 11, 1919. The engineer's preliminary estimate was $22,964.00, the contractor's bid, $22,300.50. The value of work done during the year is $5,480.00, total done to date, the same.

L. L. Hadley, Assistant Engineer, is in charge.

The 60-foot trench specified on the plans has been excavated to rock. Nine test drill holes have been drilled in the southerly end of this trench and five of them grouted.

Contract U

This contract is for repairing the manholes of the sewer in Benton creek, Seneca Falls. It was awarded to Smith Soper, being signed on January 3, 1919. Construction work began January 6, 1919. The engineer's preliminary estimate was $5,941.00, the contractor's bid, $7,382.00. The value of work done during the year is $5,147.68. The work was accepted April 16, 1919, and the final account, amounting to $5,147.68, was approved by the Canal Board August 6, 1919.

L. L. Hadley, Assistant Engineer, was in charge.

The contract work was completed on March 25, 1919.

Contract Q

This contract is for constructing pile dolphins on Cayuga and Seneca lakes. It was awarded to W. F. Martens, being signed on March 3, 1919. Construction work began April 24, 1919. The engineer's preliminary estimate was $5,225.00, the contractor's bid, $5,092.00. The value of work done during the year is $5,090.00, total done to date, the same.

L. L. Hadley, Assistant Engineer, is in charge.

The 19 dolphins called for by the contract were constructed and construction work was completed on June 30, 1919.

Concrete Dockwall at Canandaigua Lake Harbor
(Chapter 756, Laws of 1917.)

The work of constructing a concrete dockwall at Canandaigua lake harbor, in the county of Ontario, was done under a contract awarded to W. F. Martens, which was signed on May 6, 1918, the contractor's bid being $15,012.00. Work was started on August 30, 1918, and completed on April 24, 1919. The final estimate was $11,606.89.

L. L. Hadley, Assistant Engineer, was in charge.

The old dockwall was some 1,200 feet long.

On May 6, 1918, before work was started on this contract, an agreement was entered into between the contractor and the State whereby the 560 feet of new dockwall was made to begin at the northerly end of the old dockwall, which was some 1,200 feet long, and run towards the south end rather than begin at the southerly end and run northerly. All contract prices remained unchanged.

On August 31, 1918, a second supplementary agreement was entered into between the contractor and the State whereby additional amounts of excavation and backfill, second-class concrete, metal reinforcement and chipping concrete were done.

Plate Girder Bridge at Lake Street, Geneva
(Chapter 351, Laws of 1918)

The contract for this work provides for the construction of a plate girder bridge over the old Cayuga and Seneca canal at Lake street, Geneva, Ontario county.

It was awarded to E. Brown Baker, being signed on November 23, 1918, and the contractor's bid being $69,100.00. Work was started on May 3, 1919.

L. L. Hadley, Assistant Engineer, is in charge.

By a supplementary agreement dated June 11, 1919, the work of constructing the temporary highway bridge for vehicle traffic, provided for under chapter 246, Laws of 1919, was added to this contract, increasing the amount of the contract by $1,300.00.

The old highway bridge has been removed, a coffer-dam constructed for the west abutment and the greater part of the excavation for this abutment done. The temporary bridge was constructed and is now in use.

Chief of the contractor's plant is a scow, on which is a derrick with a 70-ft. boom.

Temporary Bridge at Lake Street, Geneva
(Chapter 246, Laws of 1919)

This law provided for the construction of a temporary bridge for vehicle traffic during the construction of the plate girder bridge over the old Cayuga and Seneca canal at Lake street, Geneva, being built under chapter 351, Laws of 1918. By special agreement this work was added to the contract for building the plate girder bridge.

THE FOLLOWING STATEMENTS SHOW THE NAMES, RANK AND COM-
PENSATION OF ENGINEERS EMPLOYED IN THE MIDDLE DIVISION
OF THE DEPARTM NT OF THE STATE ENGINEER AND SURVEYOR,
TOGETHER WITH INCIDENTAL EXPENSES, FOR THE FISCAL YEAR
ENDED JUNE 30, 1919.

Ordinary Repairs to Canals — Erie Canal
Chapter 151, Laws of 1918

NAME	Rank	Rate of compensation	*Services	Travel	Total
Guy Moulton	Division engineer	$4,800 per year	$1,980 00	.....	$1,980 00
E. J. Berry	Senior assistant engineer	3,300 per year	907 50	$0 85	908 35
W. S. Morris	Estimate clerk	2,100 per year	2,257 50	3 00	2,260 50
H. L. Bassett	Cashier	1,800 per year	495 00	.....	495 00
Harvey Wagner	Stenographer	1,500 per year	550 00	.....	550 00
C. W. Chase	Chauffeur	1,500 per year	412 50	.....	41? 50
M. Sheridan	Telephone operator	840 per year	115 50	.....	115 50
John Connors	Janitor	1,200 per year	228 25	.....	228 25
John Maley	Fireman	1,080 per year	101 61	.....	101 61
I. S. Badger	Assistant engineer	2,580 per year	473 00	.....	473 00
D. R. Lee	Assistant engineer	2,340 per year	7 66	2 04	9 70
C. F. Hopstein	Junior assistant engineer	1,800 per year	631 08	46 99	678 07
J. H. Forth	Junior assistant engineer	1,680 per year	148 50	.....	148 50
M. J. Chryst	Junior assistant engineer	1,680 per year	291 54	.....	291 54
E. L. Keeler	Junior assistant engineer	1,800 per year	135 54	.....	135 54
H. R. Horton	Junior assistant engineer	1,200 per year	11 79	.....	11 79
A. W Bischel	Engineering assistant	1,080 per year	49 50	.....	49 50
Gail Bowler	Engineering assistant	840 per year	77 00	.....	77 00
C. H. Norton	Laborer	2 50 per day	214 50	.....	214 50
R. D. Smith	Laborer	2 50 per day	68 75	.....	68 75
			$9,156 72	$52 88	$9,209 60

Incidental Expenses

Livery	$12 00
Stationery and printing	35 26
Postage	52 08
Telephone and telegraph	12 62
Miscellaneous	432 99
	544 95
Total	$9,754 55

Ordinary Repairs to Canals — Black River Canal
Chapter 151, Laws of 1918

NAME	Rank	Rate of compensation	*Services	Travel	Total
R. K Sheldon	Assistant engineer	$2,580 per year	$68 66	$47 17	$115 83
J. J. Ryan	Junior assistant engineer	1,800 per year	31 94	.....	31 94
J. E. Smith	Junior assistant engineer	1,800 per year	31 94	.....	31 94
A. Moosbrugger	Engineering assistant	1,080 per year	19 16	.....	19 16
P. Ryan	Boatman	3 00 per day	16 50	.....	16 50
			$168 20	$47 17	$215 37

Incidental Expenses

Livery	15 00
Total	$230 37

* Includes additional compensation of 10 per cent allowed above base rate.

Construction of Barge Canal — Erie Canal
Chapter 147, Laws of 1908, and amendatory laws

NAME	Rank	Rate of compensation	*Services	Travel	Total
Guy Moulton	Division engineer	$4,800 per year	$850 00	$132 16	$982 16
E. J Berry	Senior assistant engineer	3,300 per year	1,183 75	176 68	1,360 43
H. L. Bassett	Cashier	1,800 per year	1,290 00		1,290 00
L. J. Mulhauser	Stenographer	1,500 per year	540 00		540 00
I. S. Badger	Assistant engineer	2,580 per year	1,194 29	19 70	1,213 99
Geo. H Briggs	Assistant engineer	2,580 per year	1,998 72	71 25	2,069 97
F. B. Crocker	Assistant engineer	2,580 per year	2,751 65	279 19	3,030 84
L. Bartlett	Assistant engineer	2,580 per year	1,267 64	368 64	1,636 28
C. W. Costello	Assistant engineer	2,580 per year	1,354 50	454 55	1,809 05
W. J. Durkan	Assistant engineer	2,580 per year	197 74	1 57	199 31
H. H. Brown	Assistant engineer	2,580 per year	850 76	77 35	928 11
J. G. Palmer	Assistant engineer	2,580 per year	2,537 00	386 75	2,923 75
C. L. Bannister	Assistant engineer	2,580 per year	709 50		709 50
R. K. Sheldon	Assistant engineer	2,580 per year	601 43	223 45	824 88
G. S. Haight	Assistant engineer	2,340 per year	2,280 24		2,280 24
D. R. Lee	Assistant engineer	2,340 per year	164 45		164 45
O'Neil	Assistant engineer	2,340 per year	896 30	27 95	924 25
Saxton	Assistant engineer	2,160 per year	887 81		887 81
Hopstein	Junior assistant engineer	1,800 per year	933 16	68 45	1,001 61
E. L. Keeler	Junior assistant engineer	1,800 per year	366 78		366 78
Geo. H. Thomas	Junior assistant engineer	1,800 per year	880 92		880 92
N. R. McLoud	Junior assistant engineer	1,800 per year	919 06	285 56	1,201 62
M. H. Boigeol	Junior assistant engineer	1,800 per year	979 84		979 84
J. J. Ryan	Junior assistant engineer	1,800 per year	414 27		414 27
J. E. Smith	Junior assistant engineer	1,800 per year	101 12		101 12
J. H. Forth	Junior assistant engineer	1,680 per day	1,233 00	4 00	1,237 00
H. C. Smith	Junior assistant engineer	1,680 per year	950 11	7 90	958 01
M. J. Chryst	Junior assistant engineer	1,680 per year	332 69	75	333 44
G. L. Stillman	Junior assistant engineer	1,680 per year	399 27		399 27
F. J. Beach	Junior assistant engineer	1,560 per day	1,426 74		1,426 74
R. E. Homan	Junior assistant engineer	1,560 per year	386 42		386 42
H. A. Shafer	Junior assistant engineer	1,500 per year	377 88		377 88
J. S. Bierbardt	Junior assistant engineer	1,440 per year	459 68		459 68
D. D. Rogers	Junior assistant engineer	1,200 per year	124 19		124 19
H. R. Horton	Junior assistant engineer	1,200 per year	70 96		70 96
R. M. R. Howard	Junior assistant engineer	1,200 per year	30 52		30 52
L. A. Kavanagh	Engineering assistant	1,080 per year	701 20		701 20
A. Moosbrugger	Engineering assistant	1,080 per year	64 18		64 18
Frank Lutz	Engineering assistant	1,020 per year	783 30		783 30
Gail Bowler	Engineering assistant	900 per year	310 65		310 65
Daniel Scanlon	Engineering assistant	900 per year	61 41		61 41
Parnell Maroney	Engineering assistant	840 per year	27 32		27 32
W. H. Benson	Engineering assistant	840 per year	18 67		18 67
E. S. Niles, Jr.	Engineering assistant	840 per year	2 33		2 33
Frank Ladd	Boatman	3 00 per day	653 40	$15 53	668 93
Arthur Preston	Boatman	3 00 per day	478 50		478 50
Patrick Ryan	Boatman	3 00 per day	422 40		422 40
L. G. Hyle	Laborer	2 50 per day	88 00		88 00
Frank Brophy	Laborer	2 50 per day	750 75		750 75
G. M. Wilcox	Laborer	2 50 per day	434 50		434 50
C. Peacock	Laborer	2 50 per day	530 75		530 75
W. T. Tanner, Jr.	Laborer	2 50 per day	792 00		792 00
Harold Higgins	Laborer	2 50 per day	396 00		396 00
C. H. Norton	Laborer	2 50 per day	426 25		426 25
D. W. Traub	Laborer	2 50 per day	5 50		5 50
W. H. Benson	Laborer	2 50 per day	57 75		57 75
F. Voorhees	Laborer	2 50 per day	261 25		261 25
E. S. Niles, Jr.	Laborer	2 50 per day	236 50		236 50
C. P. Plummer	Laborer	2 50 per day	57 75		57 75
C. W. Chase	Chauffeur	1,500 per year	537 50	36 60	574 10
John Connors	Janitor	1,200 per year	110 00		110 00
C. H. Osterhout	Fireman	1,080 per year	23 22		23 22
Dan Burhans	Gage reader	120 per year	120 00		120 00
Wm. Prettie	Gage reader	120 per year	120 00		120 00
F. J. Graves	Gage reader	120 per year	10 00		10 00
Marie Brandt Brown	Gage reader	84 per year	77 00		77 00
J. R. Bixby	Gage reader	84 per year	77 00		77 00
A. H. Hoffmeister	Gage reader	84 per year	84 00		84 00
Ida C. Powell	Gage reader	84 pe year	84 00		84 00

* Includes additional compensation of 10 per cent allowed above base rate.

Construction of Barge Canal — Erie Canal — (Continued)
Chapter 147, Laws of 1903, and amendatory laws

NAME	Rank	Rate of compensation	*Services	Travel	Total
H. L. Ropes	Gage reader	$84 per year	$61 19		$61 19
W. S. Siver	Gage reader	84 per year	84 00		84 00
Mrs J. R. Hiller	Gage reader	84 per year	28 00		28 00
J. P. Patterson	Gage reader	84 per year	56 00		56 00
L. Sitterly	Gage reader	84 per year	84 00		84 00
John Phillips	Gage reader	72 per year	72 00		72 00
L. A. Withey	Gage reader	72 per year	71 00		71 00
H. W. Hoch	Gage reader	60 per year	40 00		40 00
P. A. Wade	Gage reader	60 per year	4 50		4 50
Fred Chamberlain	Gage reader	60 per year	60 00		60 00
A. R. Gates	Gage reader	60 per year	60 00		60 00
A. R. Merritt	Gage reader	60 per year	60 00		60 00
M. Smith	Gage reader	60 per year	60 00		60 00
Wm. H. Burns	Gage reader	60 per year	60 00		60 00
Mark Quimby	Gage reader	60 per year	60 00		60 00
Geo. E. Wright	Livery			$274	274 00
L. A. Withey	Livery			125	125 00
M. K. Ryan	Livery			100	100 00
M. E. Nicholson	Livery			125 00	125 00
M. J. Colvin	Livery			25 00	25 00
			$41,606 16	$3,287 03	$44,893 19

Incidental Expenses

Office rent	$215 00	
Fuel and light	359 28	
Stationery and printing	35 74	
Postage	112 42	
Telephone and telegraph	416 34	
Miscellaneous	2,028 84	
		3,167 62
Total		$48,060 81

Construction of Barge Canal — Oswego Canal
Chapter 147, Laws of 1903, and amendatory laws

NAME	Rank	Rate of compensation	*Services	Travel	Total
Guy Moulton	Division engineer	$4,800 per year	$850 00	$8 15	$858 15
E J Berry	Senior assistant engineer	3,300 per year	632 68	4 07	636 75
H. L. Bassett	Cashier	1,800 per year	150 00		150 05
Harvey Wagner	Stenographer	1,500 per year	675 00		675 00
L. J Mulhauser	Stenographer	1,500 per year	328 48		328 40
E M. Ellis	Assistant engineer	2,580 per year	849	158 01	1,007 48
H. H. Brown	Assistant engineer	2,580 per year	1,901	239 69	2,141 45
Geo. H. Haley	Assistant engineer	2,580 per year	1,395	21 47	1,416 59
W. J. Durkan	Assistant engineer	2,580 per year	83	2 26	86 16
A. G. Card	Assistant engineer	2,580 per year	22		22 88
W. S. Saxton	Assistant engineer	2,160 per year	1,442		1,442 49
N. R. McLoud	Junior assistant engineer	1,800 per year	1,015	116 09	1,132 04
M. H. Boigeol	Junior assistant engineer	1,800 per year	955		955 13
C. F. Hopstein	Junior assistant engineer	1,800 per year	207	15 33	223 26
J. E. Smith	Junior assistant engineer	1,800 per year	15	2 76	18 76
E L Keeler	Junior assistant engineer	1,800 per year	29		29 3
M. J. Chryst	Junior assistant engineer	1,680 per year	194		194 46
W. J. Bell	Junior assistant engineer	1,320 per year	648 44		648 63
R J Storm	Junior assistant engineer	1,200 per year	1,188 88		1,188 2
					71

* Includes additional compensation of 10 per cent allowed above base rate.

Construction of Barge Canal — Oswego Canal — (Continued)
Chapter 147, Laws of 1903, and amendatory laws

NAME	Rank	Rate of compensation	*Services	Travel	Total
A. Moosbrugger	Engineering assistant	$1,080 per year	$9 58		$9 58
L. B. Hotchkiss	Engineering assistant	840 per year	148 60		148 60
E. S. Niles, Jr	Engineering assistant	840 per year	25 67		25 67
Patrick Ryan	Boatman	3 00 per day	9 90		9 90
Thos. Moran	Boatman	3 00 per day	610 50		610 50
C. H. Norton	Laborer	2 50 per day	148 50		148 50
Chas. Smith	Laborer	2 50 per day	200 75		200 75
E. S. Niles, Jr	Laborer	2 50 per day	66 00		66 00
L. G. Hyle	Laborer	2 50 per day	5 50		5 50
Patrick Hickey	Laborer	2 50 per day	522 50		522 50
E. F. Allen	Laborer	2 50 per day	115 50		115 50
C. W. Chase	Chauffeur	1,500 per year	137 50	$7 85	145 35
John Connors	Janitor	1,200 per year	275 00		275 00
M. Sheridan	Telephone operator	840 per year	38 50		38 50
C. H. Osterhout	Fireman	1,080 per year	45 00		45 00
W. S. Morris	Estimate clerk	2,100 per year		2 65	2 65
B. M. Wilcox	Gage reader	60 per year	60 00		60 00
D. D. Tompkins	Gage reader	60 per year	55 00		55 00
Leon Hallenbeck	Gage reader	60 per year	42 26		42 26
Arthur Gray	Gage reader	60 per year	12 74		12 74
			$15,116 78	$578 33	$15,695 11

Incidental Expenses

Fuel and light	$121 80
Stationery and printing	2 50
Postage	128 60
Telephone and telegraph	215 32
Miscellaneous	488 41
	956 63
Total	$16,651 74

Construction of Barge Canal — Cayuga and Seneca Canal
Chapter 391, Laws of 1909, and amendatory laws

NAME	Rank	Rate of compensation	*Services	Travel	Total
Guy Moulton	Division engineer	$4,800 per year	$520 00	$19 61	$539 61
H. C. Smith	Senior assistant engineer	2,820 per year	2,966 88	160 49	3,127 37
DeWitt H. Daley	Senior assistant engineer	3,000 per year	135 73		135 73
H. C. Allen	Consulting engineer	60 per day	60 00	1 50	61 50
Harvey Wagner	Stenographer	1,500 per year	137 50		137 50
L. L. Hadley	Assistant engineer	2,340 per year	2,192 24	245 72	2,437 96
C. L. Bannister	Assistant engineer	2,580 per year	215 00		215 00
R. Sturtevant	Assistant engineer	2,580 per year	114 44	31 77	146 21
J. H. O'Donnell	Assistant engineer	2,580 per year	95 81		95 81
C. F. Hopstein	Junior assistant engineer	1,800 per year	21 82	7 76	29 58
L. B. Westfall	Junior assistant engineer	1,800 per year	42 50		42 50
H. L. Drake	Junior assistant engineer	1,500 per year	1,396 50	5 80	1,402 30
A. Moosbrugger	Engineering assistant	1,080 per year	10 61	11 39	22 00
Jos. Duffy	Engineering assistant	1,080 per year	47 90		47 90
E. F. Allen	Laborer	2 50 per day	363 00		363 00
Wm. S. Philo	Laborer	2 50 per day	863 50		863 50
Frank Waldo	Laborer	2 50 per day	145 75		145 75
C. W. Chase	Chauffeur	1,500 per year	125 00	17 65	142 65
M. Sheridan	Telephone operator	1,080 per year	115 50		115 50
T. C. McNicholas	Gage reader	84 per year	84 00		84 00
C. N. Bacon	Gage reader	60 per year	55 00		55 00

* Includes additional compensation of 10 per cent allowed above base rate.

Construction of Barge Canal—Cayuga and Seneca Canal—(Cont'd)
Chapter 391, Laws of 1909, and amendatory laws

NAME	Rank	Rate of compensation	*Services	Travel	Total
E. F. Garbus	Gage reader	$60 per year	$55 00	....	$55 00
Wm. H. Lane	Gage reader	60 per year	60 00		60 00
C. D. Martin	Gage reader	60 per year	55 00		55 00
Timothy Regan	Gage reader	60 per year	55 00		55 00
Fred Wright	Gage reader	60 per year	60 00		60 00
A. M. Smith	Livery		....	$40 00	40 00
D M. Kellogg	Livery		....	28 00	28 00
			$9,993 68	$569 69	$10,563 37

Incidental Expenses

Office rent	$192 00	
Fuel and light	8 00	
Stationery and printing	83 36	
Postage	16 40	
Telephone and telegraph	76 35	
Miscellaneous	138 14	
		514 25
Total		$11,077 62

Construction of Barge Canal Terminals
Chapter 746, Laws of 1911, and amendatory laws

NAME	Rank	Rate of compensation	*Services	Travel	Total
Guy Moulton	Division engineer	$4,800 per year	$960 00	$5 65	$965 65
E. J. Berry	Senior assistant engineer	3,300 per year	609 94	1 90	611 84
L. J. Mulhauser	Stenographer	1,500 per year	600 48	2 67	603 15
Harvey Wagner	Stenographer	1,500 per year	250 00	....	250 00
A. G. Card	Assistant engineer	2,580 per year	2,364 52	33 68	2,398 20
C. L. Bannister	Assistant engineer	2,580 per year	1,612 50	....	1,612 50
W. J. Durkan	Assistant engineer	2,580 per year	1,189 30	100 06	1,289 36
Geo. H. Haley	Assistant engineer	2,580 per year	1,378 41	25 47	1,403 88
L. Bartlett	Assistant engineer	2,580 per year	660 56	14 73	675 29
Geo. H. Briggs	Assistant engineer	,580 per year	34 68	....	34 68
I. S. Badger	Assistant engineer	,580 per year	633 21	3 77	636 98
H. H. Brown	Assistant engineer	,580 per year	13 87	....	13 87
J E. Smith	Junior assistant engineer	,800 per year	1,248 84	....	1,248 84
C. F. Hopstein	Junior assistant engineer	,800 per year	5 50	11 05	16 55
G. L. Stillman	Junior assistant engineer	,680 per year	548 21	....	548 21
Geo. H Thomas	Junior assistant engineer	,800 per year	44 00	....	44 00
C. L. Fox	Junior assistant engineer	,680 per year	321 70	....	321 70
M. J. Chryst	Junior assistant engineer	,680 per year	104 71	....	104 71
J. H. Forth	Junior assistant engineer	,560 per year	286 00	....	286 00
H. A. Shafer	Junior assistant engineer	,560 per year	266 23	....	266 23
W. J. Bell	Junior assistant engineer	,200 per year	590 09	....	590 09
A. Moosbrugger	Engineering assistant	,080 per year	821 35	....	821 35
L. A. Kavanagh	Engineering assistant	,080 per year	301 92	....	301 92
Frank Lutz	Engineering assistant	,020 per year	216 76	....	216 76
Edmund Wilcox	Engineering assistant	840 per year	82 29	4 44	86 73
E S. Niles, Jr.	Engineering assistant	840 per year	7 76	....	7 76
Gail Bowler	Engineering assistant	840 per year	7 45	....	7 45
Parnell Maroney	Engineering assistant	840 per year	2 49	....	2 49
Patrick Ryan	Boatman	3 00 per day	244 20	....	244 20
Frank Ladd	Boatman	3 00 per day	171 60	....	171 60
Thos. Moran	Boatman	3 00 per day	438 90	....	438 90
R D. Smith	Laborer	2 50 per day	792 00	....	792 00
Patrick Hickey	Laborer	2 50 per day	140 25	....	140 25

* Includes additional compensation of 10 per cent allowed above base rate.

Construction of Barge Canal Terminals — (Continued)
Chapter 746, Laws of 1911, and amendatory laws

NAME	Rank	Rate of compensation	*Services	Travel	Total
D. W. Traub	Laborer	$2 50 per day	$22 00	....	$22 00
C. W. Chase	Chauffeur	1,500 per year	400 00	$5 75	405 75
M. Sheridan	Telephone operator	840 per year	192 50	....	192 50
Catherine Donnelly	Telephone operator	840 per year	14 68	....	14 68
John Connors	Janitor	1,200 per year	55 00	....	55 00
C. H. Osterhout	Fireman	900 per year	168 39	....	168 39
			$17,802 29	$209 17	$18,011 46

Incidental Expenses		
Office rent	$70 00	
Fuel and light	122 75	
Stationery and printing	23 75	
Postage	80 89	
Telephone and telegraph	131 63	
Miscellaneous	324 71	
		753 73
Total		$18,765 19

Glen Creek Improvement
Chapter 341, Laws of 1918

NAME	Rank	Rate of compensation	*Services	Travel	Total
C. F. Hopstein	Junior assistant engineer	$1,800 per year	$29 03	$55 36	$84 39
E. L. Keeler	Junior assistant engineer	1,800 per year	14 51	....	14 51
J. J. Ryan	Junior assistant engineer	1,800 per year	14 52	....	14 52
Gail Bowler	Engineering assistant	840 per year	7 45	....	7 45
Total			$65 51	$55 36	$120 87

Construction of Dive Culvert, Rome
Chapter 346, Laws of 1918

NAME	Rank	Rate of compensation	*Services	Travel	Total
Foster B. Crocker	Assistant engineer	$2,580 per year	$23 65	....	$23 65
Daniel Scanlon	Engineering assistant	1,060 per year	9 90	....	9 90
Frank Brophy	Laborer	2 50 per day	8 25	....	8 25
			$41 80		$41 80

Incidental Expenses		
Stationery and printing	$57 69	
Miscellaneous	1 40	
		59 09
Total		$100 89

* Includes additional compensation of 10 per cent allowed above base rate.

Construction of Lake Street Bridge, Geneva
Chapter 351, Laws of 1918

NAME	Rank	Rate of compensation	*Services	Travel	Total
L. L. Hadley.........	Assistant engineer.............	$2,340 per year	$212 76	$19 51	$232 27
H. C. Smith.........	Junior assistant engineer........	1,680 per year	122 10	....	122 10
			$334 86	$19 51	$354 37
Incidental Expenses					
Stationery and printing..					71 04
Total........					$425 41

Construction of Minetto Bridge
Chapter 716, Laws of 1915

NAME	Rank	Rate of compensation	*Services	Travel	Total
E. M. Ellis...........	Assistant engineer..............	$2,580 per year	$745 22	$22 02	$767 24
I. S. Badger...........	Assistant engineer..............	2,580 per year	236 50	....	236 50
Chas. Smith..........	Laborer........................	2 50 per day	280 50	....	280 50
			$1,262 22	$22 02	$1,284 24
Incidental Expenses					
Fuel and light...				$17 60	
Livery..				18 00	
Telephone and telegraph..				44 70	
Postage...				2 28	
Miscellaneous...				4 38	
					86 96
Total..					$1,371 20

Limestone Creek Improvement
Chapter 751, Laws of 1917

NAME	Rank	Rate of compensation	*Services	Travel	Total
W. J. Durkan.........	Assistant engineer..............	$2,580 per year	$7 88	$0 40	$8 28
C. F. Hopstein........	Junior assistant engineer........	1,800 per year	5 50	40	5 90
Total..........			$13 38	$0 80	$14 18

* Includes additional compensation of 10 per cent allowed above base rate.

Canandaigua Lake Dredging
Chapter 756, Laws of 1917

NAME	Rank	Rate of compensation	*Services	Travel	Total
Guy Moulton	Division engineer	$4,800 per year		$6 14	$6 14
H. C. Smith	Senior assistant engineer	2,820 per year	$64 62	24 00	88 62
H. H. Brown	Assistant engineer	2,580 per year	25 20		25 20
L. L. Hadley	Assistant engineer	2,340 per year	147 98		147 98
I. S. Badger	Assistant engineer	2,580 per year	236 50		236 50
C. F. Hopstein	Junior assistant engineer	1,800 per year	63 39	3 17	66 56
E. L. Keeler	Junior assistant engineer	1,800 per year	4 84	18 72	23 56
H. L. Drake	Junior assistant engineer	1,680 per year	343 04	52 96	396 00
Patrick Ryan	Boatman	3 00 per day	3 30		3 30
L. G. Hyle	Laborer	2 50 per day	2 75		2 75
W. S. Morris	Estimate clerk	2,100 per year		6 14	6 14
C. W. Chase	Chauffeur	1,500 per year		1 00	1 00
Total			$891 62	$112 13	$1,003 75

Cowasselon Creek Dredging
Chapter 781, Laws of 1917

NAME	Rank	Rate of compensation	*Services	Travel	Total
E. J. Berry	Senior assistant engineer	$3,300 per year		$1 36	$1 36
W. J. Durkan	Assistant engineer	2,580 per year	$7 63		7 63
D. R. Lee	Assistant engineer	2,340 per year	217 16	27 00	244 16
E. L. Keeler	Junior assistant engineer	1,800 per year	19 35		19 35
J. J. Ryan	Junior assistant engineer	1,800 per year	51 13		51 13
C. F. Hopstein	Junior assistant engineer	1,800 per year	4 84		4 84
Patrick Ryan	Boatman	3 00 per day	23 10		23 10
L. G. Hyle	Laborer	2 50 per day	16 50		16 50
W. H. Benson	Laborer	2 50 per day	8 25		8 25
E. S. Niles, Jr.	Laborer	2 50 per day	5 50		5 50
			$353 46	$28 36	$381 82

Incidental Expenses		
Postage		$0 83
Livery		30 00
Miscellaneous		6 50
		37 33
Total		$419 1

Construction of Whitesboro Street Bridge, Rome
Chapter 753, Laws of 1917

NAME	Rank	Rate of compensation	*Services	Travel	Total
E. J. Berry	Senior assistant engineer	$3,300 per year		$4 40	$4 40
W. J. Durkan	Assistant engineer	2,580 per year	$1,026 88	141 14	1,168 02
E. L. Keeler	Junior assistant engineer	1,800 per year	9 68		9 68
C. F. Hopstein	Junior assistant engineer	1,800 per year	5 50	1 90	7 40
H. R. Horton	Junior assistant engineer	1,200 per year	62 85		62 85
Patrick Ryan	Boatman	3 00 per day	3 30		3 30
L. G. Hyle	Laborer	2 50 per day	8 25		8 25
			$1,116 46	$147 44	$1,263 90

Incidental Expenses		
Telephone and telegraph		$0 40
Miscellaneous		60 30
		60 70
Total		$1,324 60

* Includes additional compensation of 10 per cent allowed above base rate.

Blue Line Surveys — Erie Canal
Chapter 151, Laws of 1918

NAME	Rank	Rate of compensation	*Services	Travel	Total
E. J. Berry	Senior assistant engineer	$3,300 per year	$163 41	$1 50	$164 91
R. K. Sheldon	Assistant engineer	2,580 per year	2,040 99	475 15	2,516 14
C. W. Costello	Assistant engineer	,580 per year	1,419 00	13 92	1,432 92
E. M. Ellis	Assistant engineer	,580 per year	989 64	607 77	1,597 41
A. G. Card	Assistant engineer	,580 per year	386 09	120 58	506 67
C. L. Bannister	Assistant engineer	,580 per year	236 50		236 50
D. R. Lee	Assistant engineer	,340 per year	2,063 33		2,063 33
J. Otis Burt	Assistant engineer	,980 per year	298 06	142 45	440 51
C. Ballard Taylor	Expert	5 00 per day	250 00		250 00
J. J. Ryan	Junior assistant engineer	,800 per year	1,379 59		1,379 59
J. E. Smith	Junior assistant engineer	,800 per year	537 13		537 13
E. L. Keeler	Junior assistant engineer	,800 per year	574 84		574 84
G. L. Stillman	Junior assistant engineer	,680 per year	821 99		821 99
M. J. Chryst	Junior assistant engineer	,680 per year	50 74		50 74
L. H. Coit	Junior assistant engineer	,680 per year	33 37	36 12	69 49
D. B. Lynch	Junior assistant engineer	,560 per year	77 42		77 42
J. S. Bierhardt	Junior assistant engineer	,440 per year	44 36		44 36
D. D. Rogers	Junior assistant engineer	,200 per year	622 92		622 92
A. H. Betts	Junior assistant engineer	,200 per year	29 35		29 35
H. R. Horton	Junior assistant engineer	,200 per year	17 74		17 74
R. M. R Howard	Junior assistant engineer	,200 per year	61 79		61 79
A. Moosbrugger	Engineering assistant	,080 per year	263 12		263 12
L. Kavanagh	Engineering assistant	,080 per year	9 58		9 58
E. D. Pieri	Engineering assistant	840 per year	223 55		223 55
W. H Benson	Engineering assistant	840 per year	122 54		122 54
L E Jenkins	Engineering assistant	840 per year	45 16		45 16
E. M. Wilson	Engineering assistant	840 per year	108 38		108 38
E. S. Niles, Jr.	Engineering assistant	840 per year	108 68	4 44	113 12
Gail Bowler	Engineering assistant	840 per year	66 04		66 04
Parnell Maroney	Engineering assistant	840 per year	434 94		434 94
Patrick Ryan	Boatman	3 00 per day	310 20		310 20
Frank Brophy	Laborer	2 50 per day	110 00		110 00
D. W. Traub	Laborer	50 per day	68 75		68 75
C. P. Plummer	Laborer	50 per day	55 00		55 00
L. G. Hyle	Laborer	50 per day	1 25		19 25
E. S. Niles, Jr.	Laborer	50 per day	00		1 00
			$14,054 45	$1,401 93	$15,456 38

Incidental Expenses

Livery.. $982 25
Postage... 48 84
Telephone and telegraph... 6 45
Miscellaneous... 622 22

1,659 76

Total... $17,116 14

Blue Line Surveys — Oswego Canal
Chapter 151, Laws of 1918

NAME	Rank	Rate of compensation	*Services	Travel	Total
E. J. Berry	Senior assistant engineer	$3,300 per year	$53 22	$1 00	$54 22
R. K. Sheldon	Assistant engineer	2,580 per year	62 42	14 00	76 42
J J. Ryan	Junior assistant engineer	1,800 per year	43 55		43 55
R. J Storm	Junior assistant engineer	1,200 per year	14 19	3 08	17 27
E. D. Pieri	Engineering assistant	840 per year	22 35		22 35
W. H. Benson	Engineering assistant	840 per year	20 32		20 32
			$216 05	$18 08	$234 13

Incidental Expenses

Livery.. 32 00

Total... $266 13

* Includes additional compensation of 10 per cent allowed above base rate.

Surveys for State Court of Claims — Erie Canal
Chapter 151, Laws of 1918

NAME	Rank	Rate of compensation	*Services	Travel	Total
Geo. H. Briggs........	Assistant engineer..............	$2,580 per year	$501 80	....	$501 80
D. R. Lee...........	Assistant engineer..............	2,340 per year	62 90	....	62 90
Geo. H. Thomas.......	Junior assistant engineer........	1,800 per year	976 43	$135 18	1,111 61
C. F. Hopstein........	Junior assistant engineer........	1,800 per year	10 65	6 84	17 49
M. J. Chryst.........	Junior assistant engineer........	1,680 per year	9 22	....	9 22
H. R. Horton........	Junior assistant engineer........	1,200 per year	7 10	....	7 10
L. A. Kavanagh.......	Engineering assistant............	1,080 per year	125 39	....	125 39
Gail Bowler..........	Engineering assistant............	840 per year	4 97	....	4 97
			$1,698 46	$142 02	$1,840 48

Incidental Expenses

Livery...	$512 00	
Stationery and printing..........................	60	
Postage..	40 00	
Telephone and telegraph..........................	1 25	
Miscellaneous....................................	99 96	
		653 81
Total...		$2,494 29

Survey for Hospital Development Commission — Utica State Hospital, Marcy Division
Chapter 151, Laws of 1918

NAME	Rank	Rate of compensation	*Services	Travel	Total
Lewis Bartlett.........	Assistant engineer.	$2,580 per year	$845 30	$357 68	$1,202 98
George H. Thomas.....	Junior assistant engineer........	1,800 per year	33 65	....	33 65
L. H. Coit...........	Junior assistant engineer........	1,680 per year	305 80	....	305 80
G. L. Stilman........	Junior assistant engineer........	1,680 per year	27 50	....	27 50
L. Kavanagh..........	Engineering assistant.	1,080 per year	49 91	....	49 91
Frank Luts..........	Engineering assistant.	1,020 per year	28 44	....	28 44
Gail Bowler..........	Engineering assistant.	840 per year	78 13	....	78 13
C. P. Plummer........	Laborer......................	2 50 per day	66 00	....	66 00
Total.............		....	$1,434 73	$357 68	$1,792 41

* Includes additional compensation of 10 per cent allowed above base rate.

SUMMARY

The foregoing tables are summarized as follows:

Ordinary Repairs to Canals

1. Erie canal, chapter 151, Laws of 1918....................................	$9,754 55
2. Black River canal, chapter 151, Laws of 1918............................	230 37

Construction of Barge Canal

3. Erie canal, chapter 147, Laws of 1903, and amendatory laws..............	48,060 81
4. Oswego canal, chapter 147, Laws of 1903, and amendatory laws............	16,651 74
5. Cayuga and Seneca canal, chapter 391, Laws of 1909, and amendatory laws..	11,077 62

Construction of Barge Canal Terminals

6. Barge canal terminals, chapter 746, Laws of 1911, and amendatory laws.....	18,765 19

Special Work

7. Glen creek improvement, chapter 341, Laws of 1918......................	120 87
8. Dive culvert, Rome, chapter 346, Laws of 1918.........................	100 89
9. Lake street bridge, Geneva, chapter 351, Laws of 1918.................	425 41
10. Minetto bridge, chapter 716, Laws of 1915...........................	1,371 20
11. Limestone creek improvement, chapter 751, Laws of 1917..................	14 18
12. Canandaigua Lake dredging, chapter 756, Laws of 1917..................	1,003 75
13. Cowasselon creek, dredging, chapter 781, Laws of 1917..................	419 15
14. Whitesboro street bridge, Rome, chapter 753, Laws of 1917..............	1,324 60

Special Surveys

15. Blue line surveys, Erie canal, chapter 151, Laws of 1918.................	17,116 14
16. Blue line surveys, Oswego canal, chapter 151, Laws of 1918.............	266 13
17. Surveys for State Court of Claims, Erie canal, chapter 151, Laws of 1918....	2,494 29
18. Survey for Hospital Development Commission — Utica State hospital, Marcy division, chapter 151, Laws of 1918...................................	1,792 41
Total...	$130,989 30

REPORT

OF THE

DIVISION ENGINEER

OF THE

WESTERN DIVISION

For the Fiscal Year Ended June 30, 1919

[155]

WESTERN DIVISION

State of New York

Department of State Engineer and Surveyor

Western Division

Rochester, N. Y., *July* 1, 1919.

Hon. Frank M. Williams, *State Engineer and Surveyor,*
Albany, N. Y.:

Sir.—I have the honor of submitting herewith a report covering the work of the Western Division for the fiscal year ended June 30, 1919.

On December 31, Mr. F. P. Williams, Division Engineer since April 1, 1914, resigned to accept your appointment as Special Deputy State Engineer. The last half of the fiscal year the activities of the Division have been under my supervision.

About 70 per cent of the improved canal within the limits of the Western Division follows the old Erie canal. The necessity of maintaining traffic through the old canal while the enlargement was under way retarded construction, in some instances limiting it to the closed season. There were also delays in arriving at agreements for railroad crossings and in determining a location which would best fulfill the requirements of the city of Rochester. When, in 1915, it became necessary to suspend the awarding of contracts on account of the exhaustion of the original one hundred and one million dollar appropriation, there were certain sections in this Division not under contract. The interval during which the awarding of contracts for the uncompleted sections was held up delayed the opening of the improved canal through the Western Division beyond the date when it was opened through the Middle and Eastern Divisions. As soon as additional funds (provided by chapter 570, Laws of 1915) were made available, work was energetically pushed and the beginning of the fiscal year just closed found boats passing the entire length of the new canal. But, in order to open the waterway at the earliest possible time, some sections were not excavated to full width and unessential structural details were omitted. It is the completion of such work, together with the construction of Rochester harbor and terminal facilities at Buffalo and operations at various points along the improved canal, which has constituted the principal part of the work of the Division during the year.

Special appropriations for eight different projects have also required surveys or inspective supervision by employees of this department. Of these the blue line surveys have required the major portion of time. Up to the present year blue line surveys in the Western Division have received only intermittent attention, but with the prospect that men and funds would be available to push the work to completion this year it was placed in charge of a Senior Assistant Engineer and men are being assigned to the work as rapidly as they can be released from construction. Ellicott creek improvement at Tonawanda and Hertel avenue bridge over the Erie canal at Buffalo have required a small party. The Chadakoin river improvement at Jamestown, Eighteen-Mile creek culvert at Lockport and Griffin creek improvement at Cuba, have required an occasional day's assignment. Numerous surveys, maps, reports and investigations have been made in connection with claims which have been filed against the State. Two surveys in connection with applications for grants of lands under water in the Niagara river have been made.

The State-owned lands along the water-front are increasing in value each year as the hydraulic power-supply is becoming more extensive and navigation facilities are increased. Applications for grants which have been made during recent years have required surveys covering the water-front from Buffalo creek southwesterly to Lackawanna and also about three of the twenty-five miles of river-front between Buffalo creek and Niagara Falls. The method of surveying individual, isolated parcels, as grants are requested, is not as satisfactory as it would be to make a continuous survey. Much of the work consists in retracing old shore lines, and this can be done with greater accuracy if carried through long sections. The value of the land is sufficient to warrant accurate surveys in determining the bounds of that owned by the State, and if funds can be obtained, I would recommend the making of an accurate survey and map of all State lands along the Niagara river to Niagara Falls and along the lake-front adjacent to the city of Buffalo.

Investigations and reports were made on the condition of the canal channel before the opening of navigation this spring. When possible, this was done in the dry. In sections that were not unwatered, investigations were made by means of a sweep-boat and soundings. Sediment has been deposited at stream entrances and sloughing of the banks has occurred at isolated points. These have not been sufficient seriously to interfere with navigation, but frequent examinations of the channel should be made, since changes are constantly occurring. The sweep-boats used by the Department are rather crude affairs, rigged up for examining small areas, as construction work progressed. These are not adequate for examining long sections of canal, and in order that obstructions may be quickly located and the channel frequently examined, sweep-boats suitable for the work should be added to the equipment.

The canal improvement, to which the major portion of the energies and time of the engineering force has been devoted during the last score of years, has permitted a close study of engineering problems in the territory adjacent to the canals and has furnished valuable data for the intelligent consideration of engineering questions which may be submitted to the State Engineer in his capacity as engineer adviser to the State. Outside of the territory adjacent to the canals, however, there has not been an equal opportunity for the study of engineering problems. There are more or less frequent appropriations by the Legislature for the improvement of creeks and rivers, in order to promote navigation or to avoid the over-flow of banks in flood season, the destruction of valuable property or the menace to health, and also there are occasional calls from heads of State institutions or other departments for surveys and advice in regard to water-supply or sewage works, but in general the acts providing for such miscellaneous construction carry no funds for a continuous study of the problem, and as a consequence the plans for repairs, as well as for original construction, have to be based on

such data as the engineers may be able to gather in a limited
time.   The lack of adequate information is sometimes a very
serious handicap.   For example, reports on a new river improvement are occasionally requested at short notice, and in such work
the records of flood flows and studies of conditions over a period
of years are indispensable to the most efficient and economical
expenditure of funds.   The intermittent and local character of
these special appropriations does not permit a continuous study
of the requirements nor an observation of the effect of improvements already made, and as a result the report or engineering
advice from this Department cannot be so reliable nor so extensive
as might be rendered if funds were available for following up
improvements.   If construction work already done could be
inspected from time to time, to note its effectiveness, it would provide a more intelligent basis for any further expenditures which
may be contemplated.   I would therefore suggest for your consideration the advisability of procuring a small fund to be allotted
for the investigation and study of engineering problems outside
of canal territory.

For the purpose of maintaining navigation during the construction of the Rochester harbor, a temporary dam was built in the
Genesee river about 3,000 feet below the canal crossing.   A rapid
rise in the Genesee river, shortly before the opening of navigation
this spring, brought down large quantities of trees, logs and heavy
drift, which lodged against the boom placed for protection in front
of the dam, breaking the boom and passing down to the dam.
This material accumulated too rapidly to permit the lowering of
the dam and efforts were concentrated on removing the drift.
Work was carried on continuously, teams and a traction engine
being used to draw the heavy material ashore.   The unflagging
efforts of the contractor had practically cleared the face of the
dam when a second flood occurred, with greater flow.   This
brought such a pressure against the dam that a portion of the
foundation was torn from the bed rock to which it was bolted
and about one hundred feet was carried away during the night
of May 23.   Materials for repairs were assembled and as soon
as the water receded to a stage which would permit, cribs, pre-

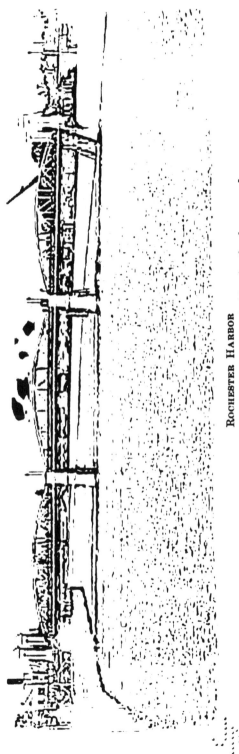

ROCHESTER HARBOR

View looking north, with new Clarissa street bridge in the foreground.

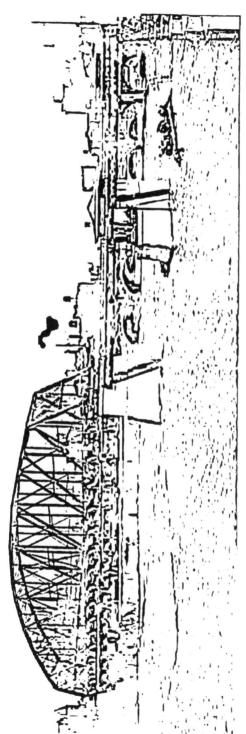

MOVABLE DAM (BRIDGE AND SECTOR GATE TYPES), ROCHESTER

A portion of the old dam is seen at the extreme left. New sections, with crests level with the old dam, replace portions of the old dam removed, so that when the movable dam is opened in the winter, exact former conditions of water-level prevail.

WEST SIDE OF GENESEE RIVER CROSSING

The channel runs through Genesee Valley Park in this vicinity and several foot-bridges, like the one in this view, were needed to connect the separated portions of the park.

WEST SHORE RAILROAD BRIDGE AT PITTSFORD

This bridge has been recently completed and also the prism excavation at this point (being completed under contract No. 179) has been recently finished.

AUBURN RAILROAD BRIDGE NEAR PITTSFORD

Recently completed. Prism excavation in this vicinity being finished under contract No. 170.

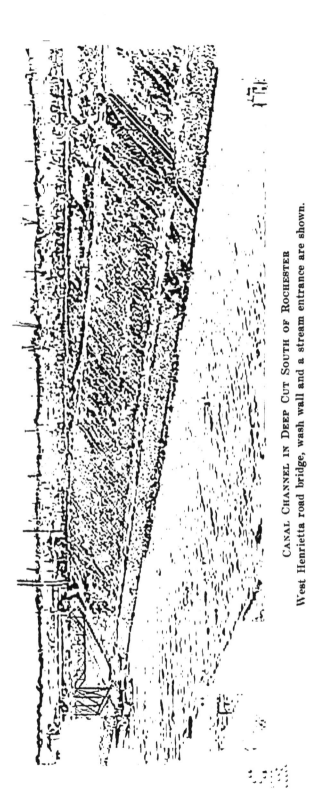

CANAL CHANNEL IN DEEP CUT SOUTH OF ROCHESTER

West Henrietta road bridge, wash wall and a stream entrance are shown.

ROCHESTER HARBOR

Construction of west river wall south of Clarissa street.

ROCHESTER HARBOR

Construction of west river wall north of Clarissa street.

viously framed, were sunk and ten days later the dam was restored. Canal traffic at the Genesee river crossing was held up a short time by the excessive current from the flood, but repairs to the dam kept pace with the recession of the flood, so that there was no delay from the failure of the dam.

Construction of Barge Canal and Terminals

At the beginning of the fiscal year, with the canal so far completed as to permit the passage of boats, there was not the imperative need for progressing the uncompleted work as an aid to war operations and contractors found it increasingly difficult to obtain materials and labor. However, the work was continued through the summer. After the signing of the armistice, conditions improved and before the end of the year contract work was progressing rapidly.

Fifteen new contracts have been added to the 30 in force at the beginning of the year. Of this total of 45 contracts, 21 were brought to a close during the year and 24 are still active. Thirteen contracts have been placed under the provisions of chapter 585, Laws of 1918. On four of the contracts cancelled under the provisions of that act, agreements have been entered into with the original contractors for the completion of portions of the construction. This Department has had a general supervision of the work.

The only parts of the canal system in the division where traffic is cut off by construction are the Rochester harbor and the Ohio basin at Buffalo.

The Court street dam in the Genesee river at Rochester will be completed before the opening of navigation next season and the Rochester harbor will be open for Barge canal traffic. A channel one hundred feet wide from the canal crossing to deep water at the Erie railroad bridge over the river will provide entrance to the harbor and terminal. Plans are under way for increasing the channel width to two hundred feet.

By the end of the season it is expected that construction at the Ohio basin, Buffalo, will have reached a stage where obstructions in the entrance will be removed and passage opened for boats of 12 feet draft.

6

Railroad Crossings

On the date of the last report the problem with reference to railroad crossings, which had given so much trouble in the past, was practically solved. The main line crossing east of Lyons, the West Shore and the Auburn crossings at Pittsford and the Erie crossing south of Rochester were just at the point of completion and were reported at that time as finished, although a little work still remained to be done on some of them. Since that time the bridge on the main line across the old abandoned channel of the Clyde river just east of Lyons has been removed. A fill was first made in the old channel under this bridge, then the girders were blocked up, the rivets cut, the superstructure removed and the track ballasted, being brought to grade by November 2. The switch tracks were next ballasted, the right-of-way fences were built and finally the work was accepted on January 22, 1919, by the maintenance department of the railroad.

At the Erie railroad crossing over the main line of the canal south of Rochester the plate girders for the bridge were unloaded from cars and placed on the abutments by September 3, 1918, while the remainder of the steel was stored on the site. The superstructure was completely assembled and riveted and the floor system painted by the middle of October, 1918. Concreting in the floor system was carried to completion, the east track laid across the bridge and ballasted and this track opened to traffic by December 24, 1918. The painting of the remainder of the bridge was done by about the middle of January, 1919. The backfill at both ends of the bridge for the westerly tracks has not been completed nor have tracks been placed across the bridge.

The Pennsylvania main line bridge west of the Genesee river was completed, tracks were laid across and ballasted, and the bridge was thrown open to traffic on August 30, 1918.

On the Rochester harbor work, along the west bank of the Genesee river, the Lehigh Valley Railroad Company has shifted much of its track to temporary locations, thus permitting the construction of the harbor walls and terminal. On the west bank the Erie Railroad Company has temporarily discontinued use of the easterly track so as to permit wall construction in close proximity to this track, which has been used by the State for construction purposes. While all expenses in this locality are borne by the State, the railroad companies have coöperated in a satisfactory manner.

At Tonawanda the construction of a bascule bridge has been in progress, most of the work on both substructure and superstructure having been done during the past year. Prior to a year ago the railroad company was engaged in building approaches, a work which involved a long detour, the scheme of railroad crossings in this locality having been rearranged so as to eliminate one bridge and carry two roads over a single new bridge.

Clarissa Street Bridge

This bridge is being constructed across the Genesee river by the city of Rochester and is of such design as to provide Barge canal clearance.

Previous to the erection of this bridge the city of Rochester and the State of New York entered into an agreement whereby the State of New York was to contribute toward the cost of the bridge, which the city of Rochester was to erect. The bridge is to be, in design and construction, wider and more costly than the State would have built. The masonry is finished and steel erection is well advanced.

Genesee Valley Canal Sewer Overflows and Wall Construction

During the year the city of Rochester completed two sewer overflows which discharge into the Rochester harbor. These sewer overflows were built in connection with the sewer in the old Genesee Valley canal, which replaced the existing sewer overflows that the city authorities deemed inadequate to meet conditions imposed by Barge canal construction.

The top section of the river walls, provided at the request and expense of the city to guard against extreme flood overflow, has been built simultaneously with the remainder of the structure.

SPECIAL APPROPRIATIONS

Ellicott Creek Improvement

(Chapter 624, Laws of 1913 ($80,000); chapter 728, Laws of 1915; chapters
181 and 760, Laws of 1917 ($55,000); chapter 85, Laws of 1918
(($25,000); chapter 644, Laws of 1919)

Contractor, J. W. Hennessy, Inc., Buffalo, N. Y.

Date of contract, April 18, 1918.

Engineer in charge, R. W. Cady, Assistant Engineer.

Aggregate of appropriations..................	$160,000.00
Engineer's preliminary estimate...............	86,803.25
Contractor's bid	86,885.30
Value of work done to June 30, 1919...........	75,780.00

The first contract under this appropriation was awarded to
F. L. Cohen of Buffalo on December 10, 1914. He completed
nearly all work of raising or reconstructing bridges, but failed
to undertake the prism excavation and his contract was canceled.
The contract for completion was let to J. W. Hennessy, Inc.,
on April 18, 1918, and included, with the channel excavation to
Barge canal depth, some retaining wall construction at the west
end of the Niagara street bridge, paving of slopes under the
bridges, laying concrete sidewalks and placing wrought-iron pipe
railing. This contract has now been completed except for
removal of snags and clearing, practically all the work having
been done during the past year. A clam-shell dredge, supplemented later by a large suction dredge, was used in excavating
the prism.

Concrete Culvert Over Eighteen-Mile Creek

(Chapters 181 and 626, Laws of 1917; chapter 644, Laws of 1919)

Contractor, Savage Construction Co., Buffalo, N. Y.

Date of contract, May 31, 1919.

Amount of appropriation	$12,500.00
Engineer's preliminary estimate	10,805.00
Contractor's bid	11,070.50

This contract, which provides for the construction of a concrete culvert over Eighteen-Mile creek eastward from Pound street in the city of Lockport, was first let on January 30, 1918, to Russell R. Ames, Inc., of Rochester, N. Y. With the exception of a little clearing he did nothing and the contract was canceled, therefore, and relet as above. Excavation was begun in late June, 1919, under this new contract.

Chadakoin River (Chautauqua Lake Outlet) Improvement
(Chapter 758, Laws of 1913; chapter 728, Laws of 1915; chapter 181, Laws of 1917; chapter 644, Laws of 1919)

Amount of appropriation, $100,000.00.

A contract for the dredging of sections of river channel at Jamestown and for the construction of a new dam with Taintor gates and raceway intake, in place of Warner's dam, was let on March 23, 1916, to Geo. L. Maltby, of Jamestown, N. Y. On June 18, 1918, this contract was suspended on account of slow progress. During the past year the Superintendent of Public Works has done such work as was necessary to preserve incomplete portions of the contract and to protect the State. Considerable excavation had been done just below Fairmount avenue bridge and the new bulkhead headgates were finished. The Taintor gates have been assembled and riveted, but operating machinery has not been placed. The gates are held open with heavy timber blocking. Water is controlled by temporary timber construction in front of the Taintor gates.

Hertel Avenue Bridge
(Chapter 761, Laws of 1917)

Contractor, Lupfer & Remick, Buffalo, N. Y.

Date of contract, March 15, 1918.

Engineer in charge, E. H. Anderson, Assistant Engineer.

Amount of appropriation $30,000.00
Engineer's preliminary estimate 27,937.50
Contractor's bid 27,967.20
Final estimate 25,311.20

Work was commenced in March, 1918, and previous to a year ago the excavation had been made for the abutments. This year the abutments were completed and the steel erected, and the contract finished on May 6, 1919.

Griffin Creek Improvement
(Chapter 565, Laws of 1918)

Appropriation, $15,000.00.

This appropriation was made to cover bank protection and removal of silt from the bed of Griffin creek in the village of Cuba so as to relieve adjoining lands from flood damage, the creek having been diverted from its natural course and used formerly as a feeder to the Genesee Valley canal. Plans were prepared for a contract to cover this work, but the Superintendent of Public Works decided to utilize available forces and plant and by this means he extended the Main street bridge to give greater waterway and made some further improvement. The survey work was done by this office, when requested.

Survey of Eighteen-Mile Creek
(Chapter 425, Laws of 1918)

Appropriation, $2,500.00.

This survey extended from the canal at Lockport to Lake Ontario and was made as required, U. S. Deep Waterway maps being utilized as far as possible. The power and property interests along the creek were investigated and hydrographic studies were made, resulting in two sets of estimates, one for improvements necessary to provide for the passage of 500 cubic feet per second of water down the creek channel and the other for the passage of 1,000 cubic feet per second.

Cuba–Olean Highway Bridge
(Chapter 637, Laws of 1919)

Appropriation, $24,000.00.

A survey covering the site of the existing bridge across the outlet channel from the Cuba reservoir has been made for the bureau of bridge design.

PRESERVATION OF OLD RECORDS

In addition to the large number of old and valuable plans on file in this office there are a few manuscript records of historical and legal value, which were in bad condition. Several of these have been typewritten and bound in volumes suitable for reference and preservation, additional copies having been deposited in the Albany office of the State Engineer. Other volumes, containing original maps or documents, have been suitably repaired and rebound. While the Division office has some vault space and a limited amount of partially fire-proof filing cases, it affords no protection to most of the plans and records, and these would probably be lost in the event of a severe fire in the building in which the headquarters are situated. It would be of distinct advantage to have the Division offices housed in a modern fire-proof structure.

The field construction work is now under the supervision of three Senior Assistant Engineers, A. E. Steere, in charge of the eastern end of the Division, B. E. Failing, in charge of the western end of the Division, and A. R. Morse, in charge of construction at the Genesee river crossing and at the Rochester harbor and terminal. Field offices have been closed at Newark and South Greece. The offices which are being maintained are located at Lyons, Pittsford, Rochester, Tonawanda and Buffalo.

During the summer of 1918 the forces of the Division were reduced because of the large number in military service. The " Honor Roll " contains forty-eight names and among these is one casualty, Mr. George H. Yerkes, Mechanic, Company "A," 3d Regiment, 27th Division, who was killed in action September 29, 1918, in the vicinity of St. Quentin during the attack which penetrated the Hindenburg line.

The depletion of the force threw extra work on those who remained. This was cheerfully undertaken and by eliminating vacations and extending hours the work was carried out without delay. Forty men have been discharged from military service and 36 have returned to their former positions in the State service.

In behalf of the employees of the Division and myself, I desire to express appreciation for the cordial support which has been uniformly extended by you and your deputies and to thank you for the personal consideration which we have received.

Detail reports of the Senior Assistant Engineers in charge of the residencies into which the Division is divided, together with tabulations showing financial statements and disbursements, are appended. In the Senior Assistant Engineers' reports will be found descriptions of the Barge canal and terminal work done during the year.

Respectfully submitted,

L. C. HULBURD,

Division Engineer.

APPENDED REPORTS—WESTERN DIVISION

BLUE LINE SURVEYS

Senior Assistant Engineer L. S. Hulburd reports:

Summary of Previous Work

On July 1, 1918, the following work in this Division had been done:

From Lyons east to the Wayne county line the surveys and maps had been completed and the " 1900 " base line had been marked with iron pipes from Lyons easterly to Waldorf bridge, about two miles east of Clyde.

In Palmyra village the field surveys and three maps were partially completed.

Through Rochester the maps had been completed and the base line monumented from the junction of the new and old canals west of Pittsford to the junction of the two canals near South Greece.

In the vicinity of Buffalo the field work had been nearly finished, but computations and mapping remained to be done.

Summary of Work during Year

On January 2, 1919, the blue line work was resumed in this Division and has continued through the fiscal year except for interruptions due to the force being temporarily transferred to other work.

A field party began work at Lyons and has worked westerly to Kent street, Palmyra, locating the blue line of the Erie canal where it forms a boundary line of the State's property and marking the points with wrought-iron rods ¾-inch square, and where possible tying these points to near-by structures, trees, poles, etc.

A small force has been engaged in the Rochester office making preliminary computations for the location of the blue line. These computations have been completed from Lyons to a point west of Macedon.

Tracings of the size of field note books were prepared in the office, showing the necessary data for locating the blue line in the field, and duplicate prints of these were furnished to the field party.

Maps on a scale of 100 feet to the inch, plotted on mounted eggshell paper, have been nearly finished from Lyons to a point two miles east of Palmyra.

The " 1900 " base line along the old tow-path was used where possible, as a line from which the blue line points were located. Where the base line had been removed the ties were made to the mounmented offset center line of the improved canal. Preparations have been made for monumenting this base line with concrete monuments.

ERIE CANAL, RESIDENCY No. 8

Senior Assistant Engineer A. E. Steere reports:

The construction work on this residency comprises the following contracts: Contracts Nos. 47-A, 84, 141, 148, 159 and 164, and part of 172, and terminal contract No. 31 and part of No. 106. In conjunction with the construction work various reports were made on claims arising on account of overflowage, seepage from canal banks, etc. A survey, plans and estimate were completed for grading a highway around lock No. 28-A, known as contract No. 198.

This residency was in charge of L. S. Hulburd, Senior Assistant Engineer, until September 5, 1918, at which time he entered the Federal military service; since that date it has been under the supervision of the writer.

Contract No. 47-A — Special Agreement

Under a resolution of the Canal Board dated March 27, 1917, the Superintendent of Public Works proceeded to complete contract No. 47-A, which was for completing the construction of the canal from the town line about five miles southeast of the village of Clyde to a point near the New York Central railroad crossing at Lyons. The final estimate, amounting to $917,880.30, was approved by the Canal Board January 29, 1919.

F. W. Madigan, Assistant Engineer, was in charge.

This contract had been substantially completed during the previous year, only a few minor features remaining to be done.

During the past year the channel was swept, stream entrances completed at the " Y " bridge and at Sta. 6492, riprap placed at Creeger's bridge abutments, backfill placed behind about 100 linear feet of the West Shore wall and drainage ditches excavated through the spoil-banks between the West Shore railroad culverts and the canal bank. These ditches were completed on August 12, and later in the year were cleaned and the sides sloped, concluding work on the contract.

Contract No. 84

This contract is for constructing portions of a viaduct over the Clyde river and railroad tracks at Clyde. It was awarded to Lupfer & Remick, being signed on March 9, 1917. Construction work began March 2, 1917. The engineer's preliminary estimate was $83,984.50, the contractor's bid, $80,661.80. The contract price as modified by alterations Nos. 1, 2, and 3 is $83,078.66.

J. A. Sloat, Junior Assistant Engineer, was in charge to November 16, 1918, F. W. Madigan, Assistant Engineer, from November 16, 1918, to date.

Under authority of chapter 585, Laws of 1918, this contract was canceled by the Canal Board on December 27, 1918. The cancelation became effective March 19, 1919, upon approval of the Canal Board, the contractor having filed a stipulation of his compliance with the terms of the law. The actual cost of the work from April 7, 1917, to March 19, 1919, was $81,944.72, and payment of balance due on this amount was authorized by the Canal Board on May 7, 1919. The final account for work done prior to April 7, 1917, amounting to $7,445.65, was approved by the Canal Board on May 7, 1919. The amount paid on extra work orders to date is $2,000.00.

An extra work order dated September 10, 1918, which provides for removing a portion of the filling back of the south abutment of the viaduct and replacing it with material satisfactory for road purposes has been completed.

During the year the erection and riveting of the bridge superstructure was completed. Also the lattice railing on the bridge and both approaches was placed and aligned. The placing and rolling of embankment for the north and south approaches were completed.

The concrete gutter and curb for the north approach were practically completed. Also the bottom course of macadam was placed and rolled on both north and south approaches. The reinforced concrete floor and sidewalk were completed during the last days of December and forms removed. Gravel surfacing was placed on the the Orchard street approach, while gravel was delivered and stored on the site for completing the approach west of the north approach. Brick, sand and crushed stone were delivered in sufficient quantities to complete the paving and the top course of the macadam.

Under chapter 585, Laws of 1918, by a resolution of the Canal Board dated December 27, 1918, authority was given the State Engineer to expend not to exceed $3,000.00 to complete the work remaining on this contract. This resolution was amended on June 25, so that the amount to be expended was increased from $3,000.00 to $3,800.00. The work coming under this order consisted in completing brick pavement and sidewalks, in placing channel lamps, macadam and gravel on approaches, in general grading and in painting the superstructure. The forces of the State Engineer's Department have finished the uncompleted work, except the painting of the superstructure and the placing of a small amount of gravel surfacing around the north approach.

Contract No. 148

This contract is for constructing the substructure, superstructure and approaches of a highway bridge across lock No. 27 and the Clyde river at Leach street, Lyons. It was awarded to Lathrop, Shea & Henwood Co., being signed on September 5, 1917. The engineer's preliminary estimate was $65,810.60, the contractor's bid, $66,986.20. The value of work done during the year is $49,170.00, total done to date, $56,440.00.

F. W. Madigan, Assistant Engineer, is in charge.

An extra work order dated September 13, 1918, provides for grading an approach from the tow-path to Leach street bridge approach; also removing a partly destroyed building.

During the year excavation was made for and concrete was placed in the north and south abutments; also concrete was placed in two arch openings that support the rocker bents on the south lock wall.

Excavation was made and embankment placed for the north and south approaches to the bridge, but at the close of the year these are in an incomplete state. A portion of the south approach adjacent to the Canandaigua outlet spillway, was undermined and washed away by the spring floods and has not as yet been restored to finished lines.

The erection of the south span of the bridge superstructure was started on April 16, reaching the lock wall on May 6. The girders, floor system, portals and wind-bracing were bolted in place by May 20. Riveting began on May 23, and was completed on June 13. At the close of the year painting was about 70 per cent completed. The lower floor timbers, nailing strips, etc., were also in place.

Contract No. 164

This contract is for completing the construction of the canal at certain locations between Lyons and Newark and for constructing a retaining dam at Macedon. It was awarded to Lathrop, Shea & Henwood Co., being signed on October 30, 1917. Construction work began November 8, 1917. The engineer's preliminary estimate was $124,313.00, the contractor's bid, $159,848.25. The contract price as modified by alteration No. 1 is $115,728.75. Excess steel sheet-piling to the value of $7,550.00, has been authorized by the Canal Board. The value of work done during the year is $22,700.00, total done to date, $104,680.00. The amount paid on extra work orders during the year is $94,961.85, total to date, $94,961.85.

W. W. Brown, Assistant Engineer, was in charge to June 1, 1919, F. W. Madigan, Assistant Engineer, since that date.

Alteration No. 1, approved by Canal Board December 4, 1918, provides for eliminating the construction of the retaining dam at Macedon, the remaining prism excavation at the West Shore and New York Central railroad crossings, and the remaining excavation and embankment near Hill's loop. It decreases the contract price by $44,119.50.

An extra work order dated December 1, 1917, provides for installing in the spillway above lock No. 28-A, a 3-ft. by 3-ft. sluice gate. The final account, amounting to $1,500.00, was approved by the Canal Board July 24, 1918.

An extra work order dated April 26, 1918, provides for furnishing and driving 50 feet of 16-foot steel piling in the vicinity of Sta. 6646. The final account, amounting to $1,834.32, was approved by the Canal Board September 24, 1918.

An extra work order dated May 24, 1919, provides for stopping the seepage and reinforcing the canal banks at certain points, and incidental work in connection therewith. The final account, amounting to $7,126.39, was approved by the Canal Board September 24, 1918.

An extra work order dated December 5, 1918, provides for placing riprap below the dam at lock No. 27 and on the slope west of lock No. 28-A, for completing excavation at Hill's loop, New York Central and West Shore crossings and for constructing a spillway at Macedon.

All of these extra work orders have been completed. The work order of December 5, 1918, was completed on June 17, 1919, work having been carried on throughout the winter months.

During the navigation season of 1918, construction work comprised grading and placing of gravel surfacing on the road approach near the Wayne County Home, work being partly finished. Settled canal embankments were raised and stone for wash wall was loosened from the old canal and cast upon the towpath. Work of completing the canal prism at the old canal crossings at Stas. 6680 and 6738 was resumed in April and completed in June. The canal bank west of lock No. 27, also the drainage ditch and canal bank west of lock No. 28-A were completed. A small amount of work on this contract remains to be done at the close of the fiscal year.

Contract No. 159

This contract is for placing embankment on the north canal bank between Newark and Palmyra and extending Ganargua creek spillway. It was awarded to I. M. Ludington's Sons, Inc., being signed on March 27, 1917. Construction work began April 9, 1917. The engineer's preliminary estimate was $30,464.00, the contractor's bid, $28,476.00. The contract price as modified by alterations Nos. 1, 2 and 3 is $43,258.50.

R. W. Cady, Assistant Engineer, was in charge to July, 1917,

H. A. Helling, Assistant Engineer, from July to October, 1917,
H. R. Topping, Junior Assistant Engineer, from October, 1917,
to April, 1918, and W. W. Brown, Assistant Engineer, since April,
1918.

Under authority of chapter 585, Laws of 1918, this contract
was canceled by the Canal Board on July 9, 1918. The cancelation became effective August 14, 1918, upon approval of the
Canal Board, the contractor having filed a stipulation of his compliance with the terms of the law. The actual cost of the work
from August 7, 1917, to August 14, 1918, was $40,002.99, and
payment of balance due on this amount was authorized by the
Canal Board on February 13, 1919.

This contract has been completed and the total payment, including extra work orders, was $40,684.26.

The work in progress on this contract during the fiscal year comprised excavation and placing of third-class riprap below the
spillway, driving a small amount of wooden and steel sheet-piling
and placing of second-class concrete and wash wall.

Contract No. 141

This contract is for constructing a new power-station at lock
No. 29, Palmyra. It was awarded to W. F. Maas & Son, being
signed on March 8, 1917. Construction work began April 2,
1917. The engineer's preliminary estimate was $41,166.50, the
contractor's bid, $41,180.75.

R. W. Cady, II. A. Helling, Assistant Engineers, and H. R.
Topping, Junior Assistant Engineer, have been in charge.

Under authority of chapter 585, Laws of 1918, this contract
was canceled by the Canal Board on July 9, 1918. The cancelation became effective August 31, 1918, upon approval of the Canal
Board, the contractor having filed a stipulation of his compliance
with the terms of the law. The actual cost of the work from April
7, 1917, to August 31, 1918, was $24,956.58, and payment of
balance due on this amount was authorized by the Canal Board on
May 7, 1919. The total payment, including extra work orders,
was $25,221.79.

No contract work was performed during the past fiscal year.

Purchase of a storehouse was authorized by the Canal Board on December 27, 1918. The final account, amounting to $265.21, was approved by the Canal Board February 13, 1919.

Contract No. 172

This contract is for furnishing and delivering barrel buoys, and lamp-posts for aids to navigation on the Seneca, Clyde, Genesee and Tonawanda rivers. It was awarded to Lupfer & Remick, being signed on March 15, 1918. The engineer's preliminary estimate was $14,853.00, the contractor's bid, $13,063.20. The contract price as modified by alteration No. 1 is $12,921.45. The work was accepted September 24, 1918, and the final account, amounting to $12,913.35, was approved by the Canal Board September 24, 1918.

J. A. Sloat, Junior Assistant Engineer, was in charge.

The portion of the contract affecting this residency provided for furnishing and delivering 23 red and 25 black lamp-posts at Lyons for the Clyde river.

Construction on this residency was completed prior to a year ago. The final estimate for work done on this residency amounted to $379.20.

Terminal Contract No. 31 — Lyons

This contract is for constructing a dockwall and an approach at Lyons. It was awarded to Lupfer & Remick, being signed on September 30, 1916. Construction work began September 27, 1916. The engineer's preliminary estimate was $57,925.00, the contractor's bid, $51,653.80.

F. W. Madigan, Assistant Engineer, was in charge.

Under authority of chapter 585, Laws of 1918, this contract was canceled by the Canal Board on August 14, 1918. The cancelation became effective December 27, 1918, on approval of the Canal Board, the contractor having filed a stipulation of his compliance with the terms of the law. The actual cost of the work from April 7, 1917, to December 27, 1918, was $44,694.54, and payment of balance due on this amount was authorized by the Canal Board on April 16, 1919. The final account for work done prior to April 7, 1917, amounting to $11,826.55, was approved by the Canal Board on April 2, 1919.

VIADUCT AT CLYDE

This viaduct, built partly by the State and partly by the railroads, crosses the canalized Clyde river, the New York Central and the West Shore tracks, and village streets. By its construction two railroad grade crossings were eliminated. The view is up the approach from the N. Y. C. R. R. station.

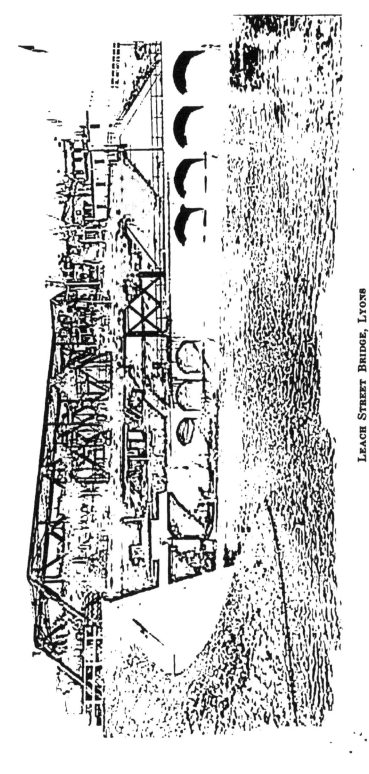

LEACH STREET BRIDGE, LYONS

This bridge spans Lock No. 27 and the adjacent stream. Here Canandaigua outlet, seen coming in at the left, and Ganargua creek unite to form the Clyde river.

MU

CANAL STRUCTURES AT LYONS

Lock No. 27 (at the extreme right), two Taintor gates and spillway, passing surplus waters of Ganargua creek, power-house, Leach street bridge and beyond it the R. S. & E. electric railway bridge.

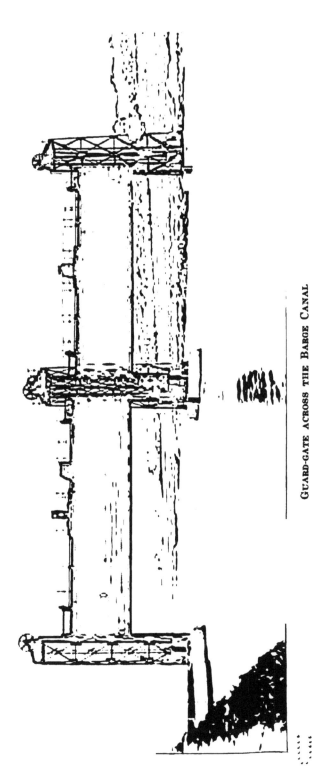

GUARD-GATE ACROSS THE BARGE CANAL.

There are several places on the canal where property would be severely damaged in case of the breaking of lock gates, the giving away of embankments, or other injury to structures, except for the presence of these guard-gates. By closing the gates a short section of canal may be cut off, thus lessening damages and allowing repairs to be made.

GUARD-LOCK JUST EAST OF GENESEE RIVER CROSSING

This lock protects the canal in time of high water in the river. The Lehigh Valley and the Erie railroad bridges are immediately beyond.

THE LEHIGH AND THE ERIE RAILROAD BRIDGES

View west from the guard-lock east of the Genesee river crossing.

This contract has been completed and the total payment, including extra work orders, was $56,521.09.

After stopping on May 18, 1918, work was resumed on July
18. Defective concrete was removed and replaced, the berme
back of the wall was graded and fenders and pipe railing placed
on the dockwall. The approach to Geneva street was completed
in October. A wooden warehouse, 32 ft. by 50 ft., was added
to the contract and completed in December. The terminal as
completed has a dockwall 360 feet long with wing walls at each
end 40 feet long.

Terminal Contract No. 106

This contract is for furnishing fourteen two-ton steam tractor
cranes for Barge canal terminals. One of these is for use at Lyons.
The contract was awarded to the John F. Byers Machine Co.,
being signed on February 14, 1918. The engineer's preliminary
estimate was $5,250.00 per crane, the contractor's bid, $5,265.00
per crane. The contract price as modified by alteration No. 1 is
$5,515.00 per crane. The value of the work done at Lyons dur-
ing the year is $5,515.00. The work was accepted September 24,
1918, and the final account, amounting to $5,515.00 for the work
at Lyons, was approved by the Canal Board October 9, 1918.

The crane for the Lyons terminal was delivered in July, 1918.

ERIE CANAL, RESIDENCY No. 9

Senior Assistant Engineer A. E. Steere reports:

The construction work on Residency No. 9 under my supervision comprises contracts Nos. 21-A, 23-A, 63-A, 161, 170, 179,
183, 190, 200 and 201, and part of No. 172. The first three mentioned, having been awarded prior to the war, were carried on
during the year under chapter 585, Laws of 1918, which provided
for completing the work at actual cost from April 7, 1917, until
canceled by resolution of the Canal Board. The remaining contracts mentioned above were under construction by the contractors' forces.

Contract No. 200 includes construction at various points

between Rochester and Lockport. A portion of this work is located within the limits of Residency No. 10, but as it was handled without reference to residency divisions, report for the full contract is made under Residency No. 9.

Plans and estimates were prepared in this office for completing the unfinished work on contracts Nos. 23-A and 63-A. The new contract numbers are 190 for 23-A, and 201 and 189 for completing 63-A and a part of old contract No. 49.

Surveys were made from the junction lock at South Greece to Long Pond on Lake Ontario, in order that a study could be made of a method for handling surplus waters from the spillway, when the old canal through the city of Rochester is abandoned.

In addition to the supervision of construction, numerous reports and investigations of a hydraulic nature were prepared for the defence of the State before the State Court of Claims.

Contract No. 63-A

This contract is for completing the construction of the canal from the west line of Wayne county to the east end of contract No. 23-A at King's Bend. Length, 12.22 miles. It was awarded to the State Highway Construction Co., being signed on February 23, 1916. Construction work began on April 6, 1916. The engineer's preliminary estimate was $567,745.70, the contractor's bid, $488,103.20. The contract price as modified by alterations Nos. 1, 2, and 3 is $581,861.30. Excess embankment to the value of $16,000.00 has been authorized by the Canal Board.

D. E. Bellows, Assistant Engineer, was in charge.

This contract was canceled by resolution of the Canal Board on March 6, 1918, and thereafter up to a certain point the work was carried on by the forces of the Superintendent of Public Works. Then the remainder was included in extra work orders under contracts Nos. 179 and 201.

The actual cost of the work from April 7, 1917, to March 6, 1918 (settlement made under chapter 585, Laws of 1918), was $152,142.28, and payment of balance due on this amount was authorized by the Canal Board on August 6, 1919. The final account for work done prior to April 7, 1917, amounting to $235,-634.96, was approved by the Canal Board on May 7, 1919.

The value of work done by the Superintendent of Public Works during the year amounted to $337,843,00, total to date, $672,868.62.

During the navigation season of 1918 dredging operations were carried on by hydraulic and dipper-dredges and by a derrick-boat equipped with a clam-shell bucket. The hydraulic dredge *Ontario II* started excavation during the week of June 14 near the Auburn railroad crossing and continued work at that location until July 27, when operations were discontinued on account of hard material. The dredge was then moved to Sta. 2000, near Bushnell's basin, and resumed operations, working there until the middle of September, cleaning a section of channel about one-quarter mile in length. From this location the dredge was moved to a point immediately east of Knapp's bridge and resumed operations, working in an easterly direction, covering a distance of 4,900 feet. On October 22 this plant was dismantled and shipped from the site.

The dipper-dredge *St. Johnsville,* the derrick-boat *Powhattan* and scows were brought onto the work during June, 1918, and assembled. Dredging operations started about July 14. The *St. Johnsville* widened the prism on the east bank south of Fairport widewaters between Stas. 1985 and 1968. It moved then to the north of the widewaters and resumed prism dredging in conjunction with the *Powhattan.* The bridge at Church street, Fairport, was reached on December 16, when operations stopped and the plant was taken to winter quarters. The prism between the limits mentioned above was left in an incomplete state.

Wash-wall stone was loaded into canal boats at Lyell avenue, Rochester, by means of a steam-shovel and car equipment and transported by canal boats to the site of the contract, where the material was deposited in storage piles and also at points along the canal bank, later to be rehandled and placed in wash-wall notch.

The approaches to Knapp's bridge and the south approach of the Cartersville bridge were completed and covered with gravel surfacing.

Contract No. 201

This contract is for completing prism lining at Cartersville and constructing a stream entrance near Knapp's bridge. It was awarded to I. M. Ludington's Sons, Inc., being signed on March 13, 1919. The engineer's preliminary estimate was $48,455.25, the contractor's bid, $42,824.75. The contract price as modified by alteration No. 1 is $54,824.75. The value of work done during the year is $46,460.00, total done to date, the same.

D. E. Bellows, Assistant Engineer, is in charge.

Alteration No. 1, approved by the Canal Board April 2, 1919, provides for removing material above grade in the canal prism between Stas. 1906 and 2000, also coffer-dams at Stas. 1884 and 1901. It increases the contract price by $12,000.00.

Construction operations were started on February 26, 1919, when several articles of plant were delivered on the site and erection of buildings commenced. Coffer-dams were built to inclose the site at Cartersville, which was unwatered by means of syphons discharging into the manhole at the foot of the north canal bank. The area near Fairport was unwatered by the same means.

During the week of March 25 a Marion revolving steam-shovel with train equipment started excavation in the bottom at the Cartersville end, while slip scrapers were used on the slope. By this means the excavation was carried on almost continuously until the opening of navigation on May 15.

The excavation in the prism west of Fairport was performed by teams and scrapers, which worked continuously, as weather permitted, until the opening of navigation.

The placing of second-class concrete was started at the stream entrance near Knapp's bridge and on the slope at Cartersville during the week of April 5, the stream entrance being completed during April, while the slope lining at Cartersville was continued until the opening of navigation, at which time about 80 per cent of the concrete had been placed. Also the concrete protection in the bottom of the prism joining the guard-gate was completed.

Work on this contract was discontinued upon the opening of navigation, to be resumed at a later date.

Contract No. 179

This contract is for completing the canal prism at the New York Central and the West Shore railroad crossings near Pittsford. It was awarded to I. M. Ludington's Sons, Inc., being signed

on November 9, 1917. The engineer's preliminary estimate was
$76,033.50, the contractor's bid, $79,712.20. The contract price
as modified by alterations Nos. 1 and 2 is $92,992.20. The value
of work done during the year is $27,030.00, total done to date,
$89,220.00. The amount paid on extra work orders during the
year is $992.00, total to date, the same.

D. E. Bellows, Assistant Engineer, is in charge.

Alteration No. 2, approved by the Canal Board March 19, 1919,
provides for removing material above grade in the canal prism
east of Sta. 2170. It increases the contract price by $8,000.00.

Work on this contract, suspended during the navigation season
of 1918, was resumed during the week ended January 10, 1919,
at which time excavation started at the Auburn railroad crossing
by a small force loading material into Koppel cars and depositing
it along the north canal bank east of guard-gate. Work at the
West Shore crossing was resumed during the week of March 7 by
a Marion shovel and car equipment, being carried on continuously
until the opening of navigation. The concrete sill east of the
guard-gate at Cartersville was completed on April 5.

The excavation work under the original contract is completed,
but the excavation carried on under alteration No. 2 was confined
to clearing the channel between the Auburn crossing and Mitchell's
bridge. This section was completed as to depth, with a minimum
width of 60 feet near Mitchell bridge. In order to complete the
channel to full dimensions it will be necessary to remove about
4,000 cubic yards of material.

Contract No. 23-A

This contract is for completing the construction of the canal
from King's Bend to the Lehigh Valley railroad crossing about
one-half mile east of the Genesee river. Length, 5.13 miles. It
was awarded to H. S. Kerbaugh, Inc., being signed on May 20,
1916. On July 3, 1917, it was assigned to the Empire Engineering Co., Inc., and this assignment was approved by the Superintendent of Public Works August 14, 1917. Construction work
began July 8, 1916. The engineer's preliminary estimate was
$651,703.10, the contractor's bid, $630,568.42. The contract
price as modified by alterations Nos. 1 and 2 is $745,672.42.

C. L. Baldwin, Assistant Engineer, was in charge.

Under authority of chapter 585, Laws of 1918, this contract was canceled by the Canal Board on July 9, 1918. The cancelation became effective December 4, 1918, upon approval of the Canal Board, the contractor having filed a stipulation of his compliance with the terms of the law. The actual cost of the work from April 7, 1917, to December 4, 1918, was $873,417.57, and payment of balance due on this amount was authorized by the Canal Board on May 7, 1919. The final account for work done prior to April 7, 1917, amounting to $106,502.42, was approved by the Canal Board on May 7, 1919.

This contract has been completed and the total payment, including extra work orders, was $984,058.21.

The final account, amounting to $496.49, on extra work order dated September 29, 1917, was approved by the Canal Board December 4, 1917.

The final account, amounting to $496.49, on extra work order dated December 31, 1917, was approved by the Canal Board December 4, 1918.

The work performed on this contract during the fiscal year, which was carried on under chapter 585, Laws of 1918, was done mainly by floating plant. The dipper-dredges *Peconic* and *Pontiac* excavated in the canal prism from a point east of Clinton avenue to the guard-lock, the material being either loaded into steel bottom-dump scows of about 600 cubic yards capacity and spoiled in deep water west of lock No. 33 or placed along the canal banks, where it was later rehandled and used in raising them.

The derrick-boat *Giant* reëxcavated material previously deposited by scows along the bottom angle and cast it into spoil-areas on the north and south banks, these areas having been formed prior to the opening of navigation by dikes thrown up by steam shovels and made for the purpose of raising low banks. This derrick-boat excavated from scow dump-ground and spoiled material on the north bank between lock No. 33 and South avenue bridge, and completed embankment back of the north wall at the guard-lock in the same manner.

Derrick-scows Nos. 208 and 275 excavated a ridge west of Clinton avenue, cleaned the prism under the highway bridges, trimmed material from prism slope that was beyond the reach of the dipper-dredges and placed embankment back of the south abutment of the guard-lock.

The intercepting drainage system and the stream entrance on the north bank were completed from the west Henrietta road to the guard-lock, where the water was taken into the canal. The canal banks at locks Nos. 32 and 33 were graded and trimmed and excavation for snubbing-posts made at locations required.

Concrete work, consisting of slope paving under the east and west Henrietta road bridges, snubbing-posts at locks Nos. 32 and 33 and counterweights for the east gates at the guard-lock, was completed.

The east gate, towers and bridge of the guard-lock were assembled, riveted and partly painted.

Plans were prepared in this office for the purpose of completing the remaining work by a new contract (let as contract No. 190).

Contract No. 190

This contract is for completing the canal from King's Bend to the Lehigh Valley railroad crossing at Rochester. It was awarded to the Empire Engineering Co., Inc., being signed on March 20, 1919. Construction work began March 17, 1919. The engineer's preliminary estimate was $284,752.50, the contractor's bid, $245,191.00. The contract price as modified by alteration No. 1 is $249,679.00. Excess 12-inch vitrified pipe, laid, to the value of $180.00 has been authorized by the Canal Board. The value of work done during the year is $65,230.00, total done to date, the same.

C. L. Baldwin, Assistant Engineer, is in charge.

Alteration No. 1, approved by the Canal Board May 7, 1919, provides for the removal of material lying above grade in the bottom of the canal between approximately Stas. 2251 and 2300. It increases the contract price by $4,488.00.

Work on this contract consisted of the completion of the wing-walls for the east and west Henrietta road bridges, completion of concrete spillways for stream entrances, and excavation for wash-wall notch and placing of wash wall.

The sill of the west gate of the guard-lock was repaired. Track rails for the east gate were adjusted and concreted into place. Also block in south lock wall was repaired.

A floating plant, consisting of the dipper-dredge *Pontiac* and derrick-boats *Giant* and *Powhattan* with tugs and canal barges, was assembled upon the opening of navigation. It is trimming slopes and cleaning the canal prism where necessary in preparation for the placing of wash wall.

Wash wall is shipped to the site of the contract by canal boats and deposited roughly into the wash-wall notch, later to be worked into final position when the water is drawn from the canal.

The intercepting ditch west from the west Henrietta road has been completed and several other ditches have been cleaned to final grades. The vitrified tile drain north along Clinton avenue was completed excepting the head-walls.

Contract No. 183

This contract is for aligning the bridge which crosses the Barge canal at the west Henrietta road, Rochester. It was awarded to the Donnell-Zane Co., being signed on September 11, 1918. The engineer's preliminary estimate was $6,850.00, the contractor's bid, $5,915.25. The value of work done during the year is $5,504.53, total done to date, the same. The work was accepted June 11, 1919, and the final account, amounting to $5,504.53, was approved by the Canal Board September 17, 1919.

C. L. Baldwin, Assistant Engineer, is in charge.

The contractor performed a small amount of work on the north and south abutments during the months of November and December, then closed down until March 24, 1919, when operations were resumed. The abutments and skew backs were completed during the month of April, and preparations made for swinging the bridge to its new alignment, which was completed on May 3, 1919, and opened for traffic.

Contract No. 161

This contract is for furnishing and delivering electric motors and certain machinery at Rochester. It was awarded to Lord Construction Co., being signed on August 3, 1917. The engineer's preliminary estimate was $5,972.00, the contractor's bid,

$6,452.00. The contract price as modified by alteration No. 1 is $15,867.35. Excess quantities to the value of $39.24 have been authorized by the Canal Board. The work was accepted April 2, 1919, and the final account, amounting to $15,750.20, was approved by the Canal Board August 20, 1919. The amount paid on extra work orders during the year is $455.02, total to date, $937.02.

Gordon Edson, Assistant Engineer, was in charge at the west lock and C. L. Baldwin, Assistant Engineer, at the east lock.

An extra work order dated June 18, 1918, provides for installing and also removing, when so directed, temporary machinery, electric wiring, etc., at the guard-locks. The final account, amounting to $455.02, was approved by the Canal Board February 13, 1919.

Work done during the fiscal year consisted of installing operating equipment for the east gate of the east lock and completing permanent wiring. Construction work was completed during the week ended January 24, 1919.

Contract No. 172

This contract is for furnishing and delivering barrel buoys and lamp-posts for aids to navigation on the Seneca, Clyde, Genesee and Tonawanda rivers. It was awarded to Lupfer & Remick, being signed on March 15, 1918. The engineer's preliminary estimate was $14,853.00, the contractor's bid, $13,063.20. The contract price as modified by alteration No. 1 is $12,921.45. The work was accepted September 24, 1918, and the final account, amounting to $12,913.35, was approved by the Canal Board September 24, 1918. The amount paid on extra work orders to date is $906.50, of which amount $49.00 applies to work on this residency.

J. S. Summers, Assistant Engineer, was in charge.

The portion of the contract affecting this residency provided for furnishing and delivering at Rochester 10 barrel buoys and 4 lamp-posts for use on the Genesee river. The buoys were delivered as originally planned, but the lamp-posts were sent to Tonawanda.

The final estimate for work done on this residency amounted to $651.90. Construction work was completed the previous year.

Contract No. 21-A

This contract is for completing the canal from about Sta. 2249, about 400 feet west of Genesee river, to about Sta. 2566 + 58, about 442 feet from the east end of contract No. 6. Length, 2.23 miles. It was awarded to Walsh Construction Co., being signed on February 16, 1916. Construction work began March 1, 1916. The engineer's preliminary estimate was $415,700.00, the contractor's bid, $384,928.69. The contract price as modified by alterations Nos. 1 and 2 is $428,475.54. Excess concrete to the value of $9,000.00 has been authorized by the Canal Board.

Gordon Edson, Assistant Engineer, was in charge.

Under authority of chapter 585, Laws of 1918, this contract was canceled by the Canal Board on July 17, 1918. The cancelation became effective December 4, 1918, upon approval of the Canal Board, the contractor having filed a stipulation of his compliance with the terms of the law. The actual cost of the work from April 7, 1917, to December 4, 1918, was $415,266.46, and payment of balance due on this amount was authorized by the Canal Board on May 21, 1919. The final account for work done prior to April 7, 1917, amounting to $205,061.14, was approved by the Canal Board on May 7, 1919.

This contract has been completed and the total payment, including extra work orders, was $631,400.85.

The final account, amounting to $8,919.58, on extra work order dated January 2, 1918, was approved by the Canal Board August 31, 1918.

The final account, amounting to $780.00, on extra work order dated January 2, 1918, was approved by the Canal Board August 14, 1918.

The removal of the rock-fill approach used for construction trains on the north side of the prism near Sta. 2522 was carried on until July 12 by means of a locomotive crane equipped with an orange-peel bucket. This crane was later supplanted by a floating plant and scows, which were used in excavating the fill under the construction tracks in the bottom of the prism. Some blasting was required to loosen the compacted material east of the B. R. & P. railroad crossing. This plant was operated from October 18 to

November 21, when operations ceased and the plant was dismantled and shipped away. Grading of the roadway at the guard-lock was completed and gravel surfacing placed thereon. Grading slopes between the guard-lock and Scottsville road was completed. The lock-tender's shelter at the guard-lock was completed July 12. All the plant was shipped away and the site cleaned up.

Contract No. 170

This contract is for constructing a junction lock and completing the canal prism excavation and incidental work at South Greece. It was awarded to Cleveland & Sons Company, being signed on November 10, 1917. The engineer's preliminary estimate was $54,800.50, the contractor's bid, $64,588.50. The contract price as modified by alteration No. 1 is $64,942.50. The value of work done during the year is $12,384.24. The work was accepted December 18, 1918, and the final account, amounting to $56,-444.24, was approved by the Canal Board April 16, 1919. The amount paid on extra work orders during the year is $3,651.51, total to date, the same.

A. S. Milinowski, Assistant Engineer, was in charge.

The final account, amounting to $3,651.51, on extra work order dated June 4, 1918, was approved by the Canal Board December 27, 1918. This order provided for constructing timber platforms to facilitate the operation of balance beams, for constructing a bridge across the lower end of the lock, for excavating pits to close up blind drains and for constructing drainage ditches.

The work on this contract remaining to be done at the beginning of the fiscal year comprised excavation and protection work for the weir outlet channel and the completion of a short stretch of canal prism westerly from the lock and spillway. The outlet channel was excavated by a traveling derrick equipped with a Page bucket. The canal prism excavation was completed by a floating plant equipped with an orange-peel bucket, the material being loaded upon flat scows, rehandled and deposited as spoil along the canal bank. Several farm drainage systems were built to relieve seepage conditions caused by canal construction at this location. The contract was completed and the site cleared up by the 15th of December, 1918.

Contract No. 200

This contract is for driving steel sheet-piling, placing concrete ling, etc., between Rochester and Lockport. It was awarded to Lupfer & Remick, being signed on February 26, 1919. Construction work began the first week of March, 1919. The engineer's preliminary estimate was $257,992.50, the contractor's bid, $180,248.50. Excess quantities to the value of $798.00 have been authorized by the Canal Board. The value of work done during the year is $112,540.00, total done to date, the same. The amount paid on extra work orders during the year is $9,269.00, total to date, the same.

A. S. Milinowski, Assistant Engineer, is in charge.

An extra work order dated March 14, 1919, provides for transporting steel sheet-piling owned by the State from various localities where it was stored to the points where it is to be driven on contract No. 200, and for cutting off battered ends and cutting to proper lengths and driving this sheet-piling, at special unit prices. Partial payment, amounting to $9,269.00, was approved by the Canal Board April 2, 1919.

Material and plant were shipped to the various points and construction work started during the first week of March. One pile-driver outfit was assembled at Maybee's and three drivers at South Greece and Cromwell's bridge. The pile-driving at Maybee's bridge was completed on March 21. This driver was moved to Holley, where the pile-driving was completed on April 18. This driver was then dismantled and sent to Albion, to complete the pile-driving at that point. Two pile-driving plants at South Greece and Cromwell's bridge were operated almost continuously until the work was completed May 9, when the plants were dismantled.

Excavation at Fancher for trimming the slope was started early in March and carried to completion by the end of the month. Forms were placed on the slope and materials for concrete delivered. Concrete work at this location was started during the week of April 4 and completed on April 18.

The work at Holley comprised excavation and wash wall at the guard-gate, placing of embankment at Tuttle's bridge and raising concrete spillway, all of which were carried out in detail and completed on May 13.

At Medina, culvert No. 96, work consisted of excavation, chipping of concrete from old culvert barrel, driving steel sheet-piles for cut-off at the ends of the protection and placing of concrete and waterproofing over the designated area, all of which were carried out in detail and completed on May 13, at which time water was turned in for the opening of navigation. The plant was removed and the site cleaned up.

Senior Assistant Engineer A. R. Morse reports:

The part of Residency No. 9 under my supervision is in the city of Rochester and embraces all of the work under construction that is connected with the development of the Rochester harbor and Rochester terminal of the Barge canal. It also embraces other work that is proposed or under construction incidental to the Barge canal crossing of the Genesee river in Genesee Valley park. The construction work is divided into the following contracts: Contracts Nos. 59, 138, 144, 191 and 192, and terminal contracts Nos. 48, 57 and 70. There are several other pieces of construction work in the residency intimately connected with the Barge canal, such as the Clarissa street bridge over the Genesee river, the Erie Railroad crossing of the Genesee river south of Clarissa street, the Erie Railroad and the Lehigh Valley Railroad crossings of the Barge canal in Genesee Valley park, and the Pennsylvania Railroad (main line) crossing of the Barge canal west of the Genesee river.

Considerable extra work devolved upon the engineering force of the residency, because of the cancelation and subsequent agreements for contracts Nos. 138 and 59. The work, however, under the agreement plan has progressed well toward completion.

Contract No. 59

This contract is for the construction of the canal from the west end of contract No. 23-A to the east end of contract No. 21-A, and the construction of Rochester harbor, between the crossing at

Genesee Valley park and a point about 400 feet south of the proposed dam near Court street bridge. Length along the canal, 0.63 mile, along the harbor, 3.25 miles. It was awarded to MacArthur Brothers Company, being signed on November 3, 1916. Construction work began January 3, 1917. The engineer's preliminary estimate was $1,675,252.86, the contractor's bid, $1,596,788.91. The contract price as modified by alterations Nos. 1 and 2 is $1,603,285.11. The amount paid on extra work orders during the year is $2,030.85, total to date, $68,054.31.

Arthur S. Whitbeck, Assistant Engineer, is in charge.

Under authority of chapter 585, Laws of 1918, this contract was canceled by the Canal Board on July 9, 1918. The cancelation became effective August 14, 1918, on approval of the Canal Board, the contractor having filed a stipulation of his compliance with the terms of the law. The actual cost of the work from April 7, 1917, to August 14, 1918, was $657,165.59, and payment of balance due on this amount was authorized by the Canal Board on May 7, 1919. The final account for work done prior to April 7, 1917, amounting to $20,475.95, was approved by the Canal Board on May 7, 1919.

The contractor is completing the work. The total payments to date, including extra work orders, are $1,563,026.00.

An extra work order dated September 24, 1917, provides for constructing a timber movable dam across the Genesee river at Elmwood avenue. The final account, amounting to $67,254.31, was approved by Canal Board November 13, 1918.

The uncompleted portion of the contract in Genesee Valley park east of the Genesee river was relet as a new contract (No. 192).

The season having been exceedingly open, a large amount of work has been done during the year both in the excavation of the river channel and in building river walls, leaving a very small percentage to be done. Thirty-nine hundred linear feet, or 63 per cent, of the river walls have been constructed.

During the month of May, 1919, the water in the Genesee river assumed flood conditions and brought down quantities of drift wood and trees, which lodged at the temporary dam in the Genesee river.. It was impossible at the time to remove some of the debris and consequently a large section of the dam was carried away. The dam was quickly repaired, however, by placing timber cribs in the breach, upon which the damaged portion was reconstructed.

Contract No. 192

This contract is for completing the canal from the east guard-lock to the Genesee river, and the work in Genesee Valley park. Length, 0.506 mile. It was awarded to Brown & Lowe Co., being signed on January 22, 1919. Construction work began February 1, 1919. The engineer's preliminary estimate was $327,525.00, the contractor's bid, $428,860.00. The value of work done during the year is $133,840.00, total done to date, the same.

Considerable progress was made in prism excavation, although the site was flooded several times by the waters of the Genesee river.

The north abutment of the west foot-bridge is complete and some work has been done on the south abutment. The spillway at the entrance of Red creek into the Barge canal channel is done. The north abutment of the east foot-bridge is completed and work on the south abutment has been started. The entire north retaining wall and the portion of the south wall between the Lehigh Valley R. R. bridge and the guard-lock has been completed.

Some progress has been made in filling the temporary channel of Red creek and in building road in Plymouth avenue and the approaches to Elmwood avenue bridge.

Contract No. 191

This contract is for excavating the canal channel in the Genesee river near Elmwood avenue bridge. Length, 0.62 mile. It was awarded to the Empire Engineering Co., Inc., being signed on January 14, 1919. Construction work began June 7, 1919. The engineer's preliminary estimate was $189,850.00, the contractor's bid, $176,170.00. The value of work done during the year is $7,360.00, total done to date, the same.

The contractors are excavating the rock and other material in the river channel by means of dredge, drill-boat, derrick-boat and scows.

Contract No. 144

This contract is for constructing two concrete bridges over Red creek in Genesee Valley park, Rochester. It was awarded to W. F. Martens & Co., Inc., being signed on June 14, 1917. Construction work began June 18, 1917. The engineer's preliminary estimate was $41,480.70, the contractor's bid, $41,258.70. The contract price as modified by alteration No. 1 is $46.208.70. The value of work done during the year is $840.00, total done to date, $6,580.00.

Alteration No. 1, approved by the Canal Board September 10, 1918, provides for the use of sheeting and bracing in the construction of the abutments of the lower bridge. It increases the contract price by $4,950.00.

With the exception of a small amount done during July, 1918, practically no work was done on the contract during the year.

On May 7, 1919, the Canal Board canceled the contract and directed the State Engineer to prepare plans and specifications for a new contract. These plans have been prepared and the new contract is known as No. 144-A.

Contract No. 138

This contract is for constructing a movable dam, bulkheads, retaining walls and incidental work, at Rochester. It was awarded to the Combined Construction Company, being signed on April 19, 1917. Construction work began on June 8, 1917. The engineer's preliminary estimate was $302,700.30, the contractor's bid, $321,115.12.

J. S. Summers, Assistant Engineer, is in charge.

Under authority of chapter 585, Laws of 1918, this contract was canceled by the Canal Board on July 9, 1918. The cancelation became effective August 14, 1918, on approval of the Canal Board, the contractor having filed a stipulation of his compliance with the terms of the law. The actual cost of the work from April 7, 1917, to August 14, 1918, was $72,908.94, and payment of balance due on this amount was authorized by the Canal Board on February 13, 1919.

The contractor is completing this work and the total payments to date, including extra work orders, are $471,893.00.

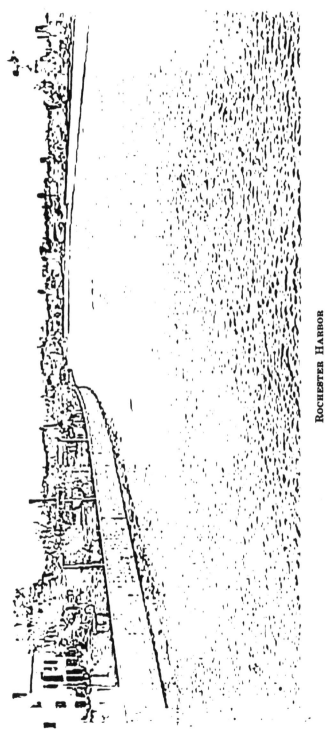

ROCHESTER HARBOR

View north from Clarissa street bridge. The terminal is seen in the distance at the right.

ROCHESTER HARBOR

View north from the east wall, showing the movable dam (bridge and sector gate (types) at the far end.

PLYMOUTH AVENUE APPROACH TO GENESEE VALLEY PARK, ROCHESTER

Progress in work of raising grade, building wall and improving road; also excavating river channel for Rochester harbor.

VIADUCT TO ROCHESTER TERMINAL

Approach to the canal terminal from Court street. As Court street is near to the business center of Rochester, probably the bulk of canal traffic will pass over this viaduct.

BASCULE BRIDGE AT MAIN AND WEBSTER STREETS, TONAWANDA AND NORTH TONAWANDA
View looking north and showing coffer-dam.

SOUTH ABUTMENT OF BASCULE BRIDGE, TONAWANDA AND NORTH TONAWANDA

In the rearrangement of railroad crossings one bridge has been eliminated and this bridge carries two roads.

BARGE CANAL TERMINAL, ERIE BASIN, BUFFALO

Construction of warehouse. The freight section, 80 by 500 feet, has a steel framework; the two-story office section, 80 by 40 feet, has a reinforced concrete frame.

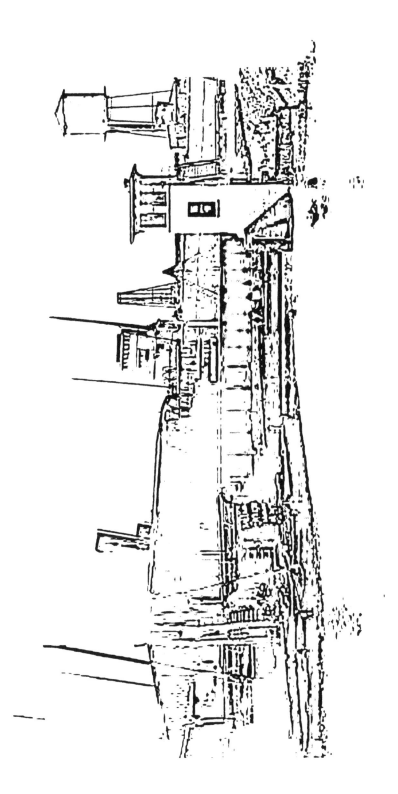

BARGE CANAL TERMINAL, OHIO BASIN, BUFFALO
New Ohio street bridge across inlet from Buffalo river to Ohio basin.

During the year the west wall and head-gates were completed as well as the east and west abutments of the bridge dam. The central pier and operator's pier were completed with the exception of the tile roofing of the operator's cabin. The sill of the bridge dam was placed early in the year, as well as the drain from the sector gate pits. The concrete for the sector gate pits is nearing completion. Excavation for the east wall is progressing and the foundation for the first section of the east head-gates is in place. Some of the steel for the sector gates was placed in position, and also two sections of the lower chord of the bridge dam were placed. The whole work originally planned is progressing well toward completion.

Terminal Contract No. 48 — Rochester

This contract is for constructing a terminal on the east side of the Genesee river at Rochester. It was awarded to Michael H. Ripton, being signed on October 19, 1916. Construction work began November 22, 1916. The engineer's preliminary estimate was $101,000.00, the contractor's bid $93,828.00. The contract has been modified by alteration No. 1, which made no change in the contract price.

C. E. Elmendorf, Assistant Engineer, was in charge.

Under authority of chapter 585, Laws of 1918, this contract was canceled by the Canal Board on July 9, 1918. The cancelation became effective August 14, 1918, on approval of the Canal Board, the contractor having filed a stipulation of his compliance with the terms of the law. The actual cost of the work from April 7, 1917, to August 14 1918, was $89,416.44, and payment of balance due on this amount was authorized by the Canal Board on March 19, 1919. The final account for work done prior to April 7, 1917, amounting to $5,350.00, was approved by the Canal Board on January 29, 1919.

This contract has been completed and the total payment including extra work orders was $94,766.00. The contract was practically completed when it was canceled.

There have been numerous track shifts by the Lehigh Valley Railroad Company, so as to accomodate construction work for the

7

terminal. The railroad company also filled the old Genesee river feeder, east of the site of the contract, and occupied this area with their tracks.

Terminal Contract No. 57 — Rochester

This contract is for constructing parts of an approach from Court street and South avenue to the terminal at Rochester. It was awarded to Charles Kiehm, being signed on February 25, 1919. Construction work began March 4, 1919. The engineer's preliminary estimate was $133,003.35, the contractor's bid, $120,597.61. The value of work done during the year is $44,820.00, total done to date, the same.

The viaduct has been completed to base of balustrade between Stas. 0 + 90.5 and 2 + 79.5. All of the columns have been placed in the canal section and on the tow-path east of the Lehigh Valley railroad tracks. Also some progress has been made on the walls and abutments specified in the contract. By the time of opening the old Erie canal through Rochester the foundations in the prism had all been completed and the work carried forward so as to cause no interference with navigation through the season.

Terminal Contract No. 70 — Rochester

This contract is for razing buildings and clearing State lands at Rochester. It was awarded to George W. Chambers, being signed on April 9, 1919. The work of demolishing buildings began on April 3, 1919. The engineer's preliminary estimate was $1,600.00 to be paid to the State of New York for all the materials that could be salvaged from the buildings and other structures on the site of the contract; the contractor's bid was $4,267.00 to be paid to the State.

There were about 43 buildings to be razed on the site of the contract and numerous fences, trees and various structures to be removed. Nearly all of these buildings have been removed above the foundation walls and some clearing of trees, fences and bushes has been done.

ERIE CANAL, RESIDENCY No. 10

Senior Assistant Engineer B. E. Failing reports:

This residency extends from the east line of Orleans county westward to the Sulphur Springs guard-lock, a distance of 43.75 miles.

All of the contracts in this residency with the exception of contracts Nos. 98 and 200, and terminal contracts Nos. 101 and 106, have been reported on as finished in the reports of previous years. A report on contract No. 98 is submitted herewith. Contract No. 200 includes construction at various points between Rochester and Lockport, extending through Residencies Nos. 9 and 10. A detailed description of the whole work will be found in report by Senior Assistant Engineer A. E. Steere for Residency No. 9. Portions of terminal contracts Nos. 101 and 106 are in both Residencies Nos 10 and 11. Final estimates on these contracts have been prepared and approved. The reports, however, are given under Residency No. 11.

Contract No. 98

This contract is for constructing a lift-bridge at Adams street and removing the existing lift-bridge at Chapel street, Lockport. It was awarded to the Tifft Construction Co., Inc., being signed on November 24, 1916. The engineer's preliminary estimate was $77,496.60, the contractor's bid, $82,276.25. The contract price as modified by alteration No. 1 is $82,426.25.

H. N. Metzger, Assistant Engineer, was in charge.

Under authority of chapter 585, Laws of 1918, this contract was canceled by the Canal Board on July 9, 1918. The cancelation became effective August 31, 1918, on approval of the Canal Board, the contractor having filed a stipulation of his compliance with the terms of the law. The actual cost of the work from April 7, 1917, to August 31, 1918, was $76,461.81, and payment of balance due on this amount was authorized by the Canal Board on February 19, 1919. The final account for work done prior to April 7, 1917, amounting to $6,919.17, was approved by the Canal Board on January 29, 1919.

This contract has been completed and the total payment, including extra work orders, was $84,723.36.

An extra work order dated April 29, 1918, provides for embankment in front of the State yards at Lockport. The final account, amounting to $842.38, was approved by the Canal Board October 9, 1918.

The work done this year consisted in finishing the approaches, building the wooden fence and placing pipe railing on the pit covers. All the other work was completed in previous years.

<hr/>

Erie Canal, Residency No. 11

Senior Assistant Engineer B. E. Failing reports:

Residency No. 11 extends from the Sulphur Springs guard-lock to and through the city of Buffalo. Contracts Nos. 19-A, 83 and 147, and part of No. 172, and terminal contracts Nos. 21, 21-P, 53, 61, 62, 66, 67, 68, 69, 107, 212 and 216 and parts of Nos. 101, 106 and 113 have all been active on this residency during the past year. Reports on these contracts follow. Also studies have been made for the track connections at the lower-town terminal, Lockport.

Contract No. 19-A

This contract is for dredging, etc., on contract No. 19, from the guard-lock at Sulphur Springs to Tonawanda. It was awarded to H. S. Kerbaugh, Inc., being signed on November 3, 1916. On July 3, 1917, it was assigned to the Empire Engineering Co., Inc., and this assignment was approved by the Superintendent of Public Works August 14, 1917. Construction work began in May, 1917. The engineer's preliminary estimate was $152,200.00, the contractor's bid, $169,750.10.

R. W. Cady, Assistant Engineer, was in charge.

Under authority of chapter 585, Laws of 1918, this contract was canceled by the Canal Board on July 9, 1918. The cancelation became effective October 9, 1918, on approval of the Canal Board, the contractor having filed a stipulation of his compliance with the terms of the law. The actual cost of the work from April 7, 1917, to October 9, 1918, was $236,943.42, and payment of balance due on this amount was authorized by the Canal Board on May 7, 1919. No work was done prior to April 7, 1917.

This contract has been completed and the total payment, including extra work orders, was $240,923.81.

Work was completed on September 6, 1918. Due to the fact that the side slopes along Tonawanda creek were unstable, slides of considerable magnitude occurred along the contract. Some of them amounted to fifty thousand yards and completely blocked the channel, thereby delaying the completion of the work. The work this year has consisted in removing the slides, cleaning up the channel, building dykes for hydraulic spoil, repairing roads which were damaged by the slides and finishing the new road at Pendleton.

Contract No. 83

This contract is for completing the canal at Tonawanda and removing a guard-lock and coffer-dam near Sulphur Springs. It was awarded to the Mohawk Dredge & Dock Co., being signed on October 22, 1917. The engineer's preliminary estimate was $149,-604.50, the contractor's bid, $216,915.00. The contract price as modified by alterations Nos. 1 and 2 is $195,351.00. The value of work done during the year is $103,266.00. The work was accepted June 25, 1919, and the final account, amounting to $158,466.60, was approved by the Canal Board July 16, 1919. The amount paid on extra work orders during the year is $3,-023.75, total to date $5,723.75.

R. W. Cady, Assistant Engineer, was in charge.

Alteration No. 2, approved by the Canal Board December 10, 1918, provides for eliminating the retaining wall on the south side of the canal easterly from the south pier of Main and Webster street bridge. It decreases the contract price by $21,264.00.

An extra work order dated July 17, 1918, provides for making fill east of the old dam and placing stone protection thereon and for pulling old piles and removing old head-gate from the channel at Tonawanda. The final account, amounting to $2,000.00, was approved by the Canal Board December 18, 1918.

An extra work order dated October 24, 1918, provides for repairing the highway east of New Home bridge. The final account, amounting to $1,023.75, was approved by the Canal Board December 4, 1918.

The work this year has consisted in completing the 12-foot channel at Tonawanda, and finishing the removal of the old guard-lock at Pendleton. The work of excavating was done by three derrick-boats, operating clam-shell buckets, and one dipper-dredge. The material was spoiled in the Government dumping grounds in Niagara river and in the back channels of Tonawanda creek. Contract work was completed on April 10, 1919.

Contract No. 147

This contract is for constructing the substructure, superstructure and approaches of a bascule bridge across Tonawanda creek at Main and Webster streets, Tonawanda and North Tonawanda. It was awarded to Scherzer Rolling Lift Bridge Co., being signed on September 10, 1917. On June 5, 1918, it was assigned to Lathrop, Shea & Henwood Co., and this assignment was approved by the Superintendent of Public Works July 1, 1918. Construction work began in October, 1917. The engineer's preliminary estimate was $227,032.80, the contractor's bid, $233,986.30. The contract price as modified by alteration No. 1 is $234,260.40. The value of work done during the year is $47,640, total done to date $78,170. The amount paid on extra work orders during the year is $2,213.56, total to date $5,858.06.

R. W. Cady, Assistant Engineer, is in charge.

Alteration No. 1, approved by the Canal Board February 19, 1919, provides for constructing a retaining wall at the southeast corner of south abutment, and for modifying the foundation of the retaining wall at the southwest corner. It increases the contract price by $274.10.

An extra work order dated July 17, 1918, provides for driving fender piles at the temporary bridge and maintaining a sidewalk at the bank building. The final account, amounting to $1,337.33, was approved by the Canal Board October 16, 1918.

An extra work order dated February 5, 1919, provides for furnishing and driving foundation piles for retaining walls, south abutment.

An extra work order dated April 7, 1919, also provides for driving fender piles at the temporary bridge and maintaining a sidewalk at the bank building.

During the year the north abutment has been finished and the coffer-dam removed; a coffer-dam has been constructed and the south abutment built; also a steel sheet-piling coffer-dam has been constructed for the center pier and the work is nearly ready for concreting. Two cars of steel for the superstructure have been received and unloaded.

Contract No. 172

This contract is for furnishing and delivering barrel buoys and lamp-posts for aids to navigation on the Seneca, Clyde, Genesee and Tonawanda rivers. It was awarded to Lufper & Remick, being signed on March 15, 1918. The engineer's preliminary estimate was $14,853.00, the contractor's bid, $13,063.20. The contract price as modified by alteration No. 1 is $12,921.45. The work was accepted September 24, 1918, and the final account, amounting to $12,913.35, was approved by the Canal Board Stepember 24, 1918. The amount paid on extra work orders to date is $906.50, of which amount $392.00 applies to work on this residency.

R. W. Cady, Assistant Engineer, was in charge.

The portion of the contract affecting this residency provided for furnishing and delivering 40 red and 40 black barrel buoys and 15 red and 15 black lamp-posts at Tonawanda. These were delivered and also four additional lamp-posts originally planned for delivery at Rochester.

The final estimate of work done on this residency amounted to $5,483.80. Work was completed the previous year.

Terminal Contract No. 101

This contract is for furnishing and installing steel stiff-leg derricks on terminal sites at Albany, Whitehall, Little Falls, Rome, Lockport and Tonawanda. This report relates to the derricks at Lockport and Tonawanda. The contract was awarded to E. Brown Baker, being signed on December 18, 1916. On February 21, 1917, it was assigned to the Mohawk Dredge & Dock Co., Inc., and this assignment was approved by the Superintendent of Public Works March 26, 1917. The engineer's preliminary estimate was $6,867.30 for these two derricks at Lockport and Tonawanda, the

contractor's bid, $10,169.30. Excess metal to the value of
$1,932.00 has been authorized by the Canal Board. The work
was accepted December 4, 1918, and the final account, amount'ng
to $11,263.22 for Lockport and Tonawanda, was approved by the
Canal Board December 27, 1918.

The final estimate was prepared during this year, contract work
having been completed the previous year.

Terminal Contract No. 21 — Erie Basin, Buffalo

This contract is for constructing a terminal at Erie basin, Buffalo. It was awarded to H. S. Kerbaugh, Inc., being signed on
January 12, 1914. It was assigned to the Empire Engineering
Co., Inc., and this assignment was approved by the Superintendent
of Public Works August 10, 1917. Construction work began in
April, 1914. The engineer's preliminary estimate was $1,513,-
925.00, the contractor's bid, $798,605.80. The contract price as
modified by alterations Nos. 1 and 2 is $797,772.30.

Elias H. Anderson, Assistant Engineer, is in charge.

Under authority of chapter 585, Laws of 1918, this contract was
canceled by the Canal Board on March 19, 1919. The cancelation
became effective April 2, 1919, on approval of the Canal Board,
the contractor having filed a stipulation of his compliance with the
law. The actual cost of the work from March 1, 1918, to April 2,
1919, was $188,963.08 and payment of balance due on this amount
was authorized by the Canal Board on June 11, 1919. The final
account for work done prior to March 1, 1918, amounting to
$645,534.37 was approved by the Canal Board on June 11, 1919.

The contractor is completing the work and the total payments to
date, including extra work orders, are $898,903.56.

An extra work order dated March 29, 1918, provides for constructing a sewer and manhole under the permanent pavement, for
placing 100 feet of conduit for electric wires at each pier and
for installing a water-main. The final account, amounting to
$1,029.11, was approved by the Canal Board August 31, 1918.

This contract was completed a year ago except for excavation in
the harbor. The contract provides for excavating, to a depth of 23
feet at mean lake level, the entire area of Erie basin between the
New York State breakwater and the shore line, a width of about

900 feet and extending from the 23-foot channel of the U. S. Government Black Rock harbor improvement southerly about 2,200 feet to the 23-foot entrance of the Buffalo river.

Exclusive outside of the winter months January, February and March, a drill-boat has worked three shifts during the year and has finished the drilling and blasting of the rock which lies above the 20-foot depth. It was removed from the contract on June 27, 1919. The earth and rock have been practically removed to a 20-foot depth. It is now the intention to complete this excavation to the 20-foot depth and discontinue work. The quantities of material excavated during the year amounted to about 35,473 cu. yds. of rock and 32,313 cu. yds. of earth.

Also a rock mound was built to protect the south end of the State breakwater.

Terminal Contract No. 21-P — Erie Basin, Buffalo

This contract is for paving part of the terminal site at Erie basin, Buffalo. It was awarded to Henry P. Burgard Company, being signed on May 6, 1918. Construction work began in May, 1918. The engineer's preliminary estimate was $14,180.00, the contractor's bid, $14,350.00. The value of work done during the year is $12,696.00. The work was accepted November 13, 1918, and the final account, amounting to $13,066.00, was approved by the Canal Board November 13, 1918.

Elias H. Anderson, Assistant Engineer, was in charge.

All of the work on this contract, with the exception of the grading, was performed during this year. It was completed in October.

Terminal Contract No. 61 — Erie Basin, Buffalo

This contract is for constructing railroad track approach to pier No. 1, Erie basin, Buffalo. It was awarded to the Walsh Construction Co., being signed on May 15, 1918. Construction work began in May, 1918. The engineer's preliminary estimate was $9,720.00, the contractor's bid, $11,650.00. The value of work done during the year is $8,960, total done to date, $10,430.

Elias H. Anderson, Assistant Engineer, is in charge.

The work this year has consisted in placing ballast and laying tracks and finishing about 75 per cent of the surface. After the rails were laid the N. Y. C. railroad refused to connect this siding with its main line. During the latter part of the year, however, a connection was made, enabling the contractor to deliver the ballast for completing his work.

Terminal Contract No. 62 — Erie Basin, Buffalo

This contract is for constructing railroad tracks and crane rails on pier No. 1, Erie basin, Buffalo. It was awarded to the Walsh Construction Co., being signed on May 15, 1918. The engineer's preliminary estimate was $8,470.00, the contractor's bid, $11,-400.00. The value of work done during the year is $9,420, total done to date, the same.

Elias H. Anderson, Assistant Engineer, is in charge.

The work of laying track was commenced in October, 1918, and the contract was completed in June, 1919, with the exception of finishing the surfacing of tracks. Work on this contract was delayed on account of the contractor waiting for a connection to be made with the tracks of the N. Y. C. railroad.

Terminal Contract No. 66 — Erie Basin, Buffalo

This contract is for placing riprap along the shore of Erie basin, between slip No. 2 and Lake street, Buffalo. It was awarded to the Empire Engineering Co., Inc., being signed on June 29, 1918. The engineer's preliminary estimate was $11,850.00, the contractor's bid, $12,820.00. The value of work done during the year is $11,400.00, total done to date, the same.

Elias H. Anderson, Assistant Engineer, is in charge.

The contractor began excavating with a dipper-dredge on September 11, 1918, completing this work in December. Riprap has been placed and the contract finished with the exception of about two scow loads of stone.

Terminal Contract No. 67 — Erie Basin, Buffalo

This contract is for constructing a railroad track approach to pier No. 2, Erie basin, Buffalo. It was awarded to the Walsh Construction Co., being signed on July 3, 1918. Construction work began in September, 1918. The engineer's preliminary estimate was $7,000.00, the contractor's bid, $7,616.00. The value of work done during the year is $7,160.00, total done to date, the same.

Elias H. Anderson, Assistant Engineer, is in charge.

All the work on the contract has been completed with the exception of the surfacing. The work was delayed on account of the contractor waiting for this track to be connected with the N. Y. C. railroad.

Terminal Contract No. 68 — Erie Basin, Buffalo

This contract is for constructing railroad track on pier No. 2, Erie basin, Buffalo. It was awarded to the Walsh Construction Co., being signed on July 3, 1918. Construction work began in September, 1918. The engineer's preliminary estimate was $6,820.00, the contractor's bid, $7,445.00. The value of work done during the year is $4,630.00, total done to date, the same.

Elias H. Anderson, Assistant Engineer, is in charge.

Work on the contract has been completed with the exception of spreading stone screenings for the surfacing. The work was delayed on account of the contractor waiting for a connection to be made with the N. Y. C. railroad.

Terminal Contract No. 69 — Erie Basin, Buffalo

This contract is for protecting the shore of Erie basin, between Lake street and slip No. 1, Buffalo. It was awarded to Richard C. Bush, being signed on February 27, 1919. Construction work began May 9, 1919. The engineer's preliminary estimate was $6,780.00, the contractor's bid, $5,886.00. The value of work done during the year is $2,140.00, total done to date, the same.

Elias H. Anderson, Assistant Engineer, is in charge.

The old crib which was to be used for the foundation of the new work was found to be in poor condition and additional crib-work was made necessary. All the additional cribwork, together with the timber cribs called for in the orginal contract were completed during the year, and the foundation was made ready for the concrete dockwall.

Terminal Contract No. 107 — Erie Basin, Buffalo

This contract is for installing electric wiring, lighting, power and battery-charging equipment for the canal terminal at Erie basin, Buffalo. It was awarded to J. Livingston & Co., Inc., being signed on March 4, 1919. The engineer's preliminary estimate was $35,025.00, the contractor's bid, $28,238.50.

No construction work has yet been done.

Terminal Contract No. 212 — Erie Basin, Buffalo

This contract is for constructing a terminal freight-house on pier No. 1, Erie basin, Buffalo. It was awarded to the Felton Construction Corporation, being signed on November 14, 1918. The engineer's preliminary estimate was $175,000.00, the contractor's bid, $182,182.00. The contract price as modified by alteration No. 1 is $181,669.00. The value of work done during the year is $15,510.00, total done to date, the same.

Elias H. Anderson, Assistant Engineer, is in charge.

Alteration No. 1, approved by the Canal Board June 25, 1919, provides for steel trusses and columns, a new type of crane rail girder, cast-iron conductor pipes with copper flashings and connections, and revised inscriptions. It decreases the contract price by $513.00.

An extra work order dated April 4, 1919, provides for installing water-service pipes, sewer pipes and a settling basin.

The freight-house is to be constructed of steel reinforced concrete and brick, and will be 500 feet long by 80 feet wide. It will have track connections on both sides of the pier.

The contractor began excavating on February 24, 1919, with a Keystone excavator and completed the excavation for the foundation in March. Foundation piles were driven and foundation walls completed in June, 1919. One thousand feet of tile drain has been laid, forms have been built for reinforced concrete columns and slab for the head-house, and reinforcing steel is being placed.

Terminal Contract No. 216 — Erie Basin, Buffalo

This contract is for constructing a frame freight-house at Erie basin, Buffalo. It was awarded to the Savage Construction Co., being signed on July 9, 1918. Construction work began in July, 1918. The engineer's preliminary estimate was $10,000.00, the contractor's bid, $9,899.00. Excess quantities to the value of $295.00 have been authorized by the Canal Board. The value of work done during the year is $10,116.00 The work was accepted November 13, 1918, and the final account, amounting to $10,-116.00, was approved by the Canal Board November 13, 1918. The amount paid on extra work orders, to date is $1,949.00.

Elias H. Anderson, Assistant Engineer, was in charge.

An extra work order dated October 24, 1918, provides for doing certain painting and electrical work and building sway braces, runways and office partition. The final account, amounting to $1,949.00, was approved by the Canal Board February 26, 1919.

This freight-house is 200 feet long by 32 feet wide and has track connections on the north side. Contract work was completed in October, 1918.

Terminal Contract No. 113

This contract is for furnishing electric capstans and trolley hoists at various canal terminals. Four capstans and one trolley hoist are to be delivered at Erie basin, Buffalo. It was awarded to the General Electric Co., being signed on June 9, 1919. The engineer's preliminary estimate was $15,000.00, the contractor's bid, $14,090.00.

No machinery has yet been delivered.

Terminal Contract No. 53 — Ohio Basin, Buffalo

This contract is for constructing a terminal at Ohio basin, Buffalo. It was awarded to the Walsh Construction Co., being signed on October 27, 1916. Construction work began in June, 1917. The engineer's preliminary estimate was $571,800.00, the contractor's bid, $532,584.00. The contract price as modified by alteration No. 1 is $597,984.00.

Elwin G. Speyer, Assistant Engineer, is in charge.

Under authority of chapter 585, Laws of 1918, this contract was canceled by the Canal Board on July 24, 1918. The cancelation became effective October 9, 1918, on approval of the Canal Board, the contractor having filed a stipulation of his compliance with the law. The actual cost of the work from April 7, 1917, to October 9, 1918, was $122,731.84 and payment of balance due on this amount was authorized by the Canal Board on May 21, 1919. No work was done prior to April 7, 1917.

The contractor is completing the work and the total payments to date, including extra work orders, are $387,512.84.

Construction work, which had been suspended on December 14, 1917, owing to war conditions, was resumed in August, 1918. During the year the retaining wall along the south end of the basin was completed, the abutments for the bascule bridge were completed, with the exception of the concrete slab on the north abutment, and foundations for the operator's cabin and two sections of wall adjacent to the abutment on the north side of the slip were completed. Due to a strike the contract was shut down during the month of May. Steel to be embedded in concrete, steel floor-beams, ladders, brackets and a safety gate have been delivered on the site. The amount of work to be done under this contract has been limited to the completion of dockwall along the south shore of the basin, the construction of seven sections of wall along Dead creek entrance and the building of the bascule bridge.

Terminal Contract No. 106

This contract is for furnishing fourteen two-ton steam tractor cranes for Barge canal terminals. The contract was awarded to John F. Byers Machine Co., being signed on February 14, 1918. The engineer's preliminary estimate was $5,250.00 per crane, the contractor's bid, $5,265.00 per crane. The contract price as modified by alteration No. 1' is $5,515.00 per crane. The value of the work done at Buffalo during the year is $5,515.00. The work was accepted September 24, 1918, and the final account, amounting to $22,060.00, Buffalo, Lockport and Tonawanda, was approved by the Canal Board October 9, 1918.

Elias H. Anderson, Assistant Engineer, was in charge.

Last year one crane was delivered at Buffalo, one at Tonawanda and one at Lockport. This year another crane has been delivered at Buffalo, completing the contract relative to the Western Division.

THE FOLLOWING STATEMENTS SHOW THE NAME, RANK AND COM-
PENSATION OF ENGINEERS EMPLOYED IN THE WESTERN DIVISION
OF THE DEPARTMENT OF THE STATE ENGINEER AND SURVEYOR,
TOGETHER WITH INCIDENTAL EXPENSES FOR THE FISCAL YEAR
ENDED JUNE 30, 1919.

Ordinary Repairs to Canals — Erie Canal

Chapter 151, Laws of 1918

NAME	Rank	Rate of compensation	*Services	Travel	Total
F. P. Williams	Division engineer	$4,800 per year	$2,520 00	$495 12	$3,015 12
L. C. Hulburd	Division engineer	4,800 per year	2,640 00	160 92	2,800 92
Waldo G. Wildes	Senior assistant engineer	3,300 per year	2,508 28		2,508 28
Anna M. Lorscheider	Stenographer	1,500 per year	1,612 50		1,612 50
E. Quans	Office assistant	1,020 per year	59 22		59 22
			$9,340 00	$656 04	$9,996 04

Incidental Expenses

Postage		$0 34
Telephone and telegraph		2 46
Miscellaneous		1 16
		$3 96
Total		$10,000 00

Construction of Barge Canal — Erie Canal

Chapter 147, Laws of 1903, and amendatory laws

NAME	Rank	Rate of compensation	*Services	Travel	Total
L. C. Hulburd	Division engineer	$4,800 per year		$63 62	$63 62
Edward Anderberg	Senior assistant engineer	3,300 per year	$20 17	41 78	61 95
B. E. Failing	Senior assistant engineer	3,540 per year	1,784 75	96 80	1,881 55
L. S. Hulburd	Senior assistant engineer	3,300 per year	907 50	9 49	916 99
A. R. Morse	Senior assistant engineer	3,300 per year	3,300 50		3,300 50
A. E. Steere	Senior assistant engineer	3,300 per year	3,508 47	158 54	3,667 01
Waldo G. Wildes	Senior assistant engineer	3,300 per year	1,029 22	43 47	1,082 69
Lewis A. Keil	Cashier	2,100 per year	2,257 50	18 36	2,275 86
Frank V. Searls	Estimate clerk	1,920 per year	1,623 39		1,623 39
W. D. Gartland	Stenographer	1,320 per year	758 59		758 59
Mary MacArthur	Stenographer	1,200 per year	330 00		330 00
M. Agnes Maloney	Stenographer	900 per year	247 50		247 50
Elizabeth S. White	Stenographer	1,200 per year	1,120 05		1,120 05
C. L. Baldwin	Assistant engineer	2,580 per year	2,773 50	146 83	2,920 33
D. E. Bellows	Assistant engineer	2,580 per year	2,773 50	956 58	3,730 08
W. W. Brown	Assistant engineer	2,160 per year	2,229 10	409 67	2,638 77
R. W. Cady	Assistant engineer	2,580 per year	2,079 77	380 40	2,460 17
R. D. Cameron	Assistant engineer	2,340 per year	1,828 05		1,828 05
C. R. De Graff	Assistant engineer	2,580 per year	668 35		668 35
Gordon Edson	Assistant engineer	2,580 per year	2,389 51	44 75	2,434 26
C. E. Elmendorf	Assistant engineer	2,580 per year	653 59	6 80	660 39
L. G. Fisher	Assistant engineer	2,340 per year	1,689 12	168 13	1,857 25
F. W. Madigan	Assistant engineer	2,580 per year	2,657 28	191 54	2,848 82

* Includes additional compensation of 10 per cent allowed above base rate.

Construction of Barge Canal — Erie Canal — (Continued)
Chapter 147, Laws of 1903, and amendatory laws

NAME	Rank	Rate of compensation	*Services	Travel	Total
Frank T Marsh	Assistant engineer	$2,340 per year	$2,273 32	$11 76	$2,285 08
H. N. Metzger	Assistant engineer	2,340 per year	675 78	69 86	745 64
A. S. Milinowski	Assistant engineer	,580 per year	2,773 50	577 30	3,350 80
Lester P. Slade	Assistant engineer	,340 per year	2,164 19		2,164 19
J. Seward Summers	Assistant engineers	,580 per year	2,284 48	45 96	2,330 44
Arthur S. Whitbeck	Assistant engineer	2,580 per year	2,773 50	72 12	2,845 62
R W. Anderson	Junior assistant engineer	1,560 per year	1,483 99		1,483 99
E. C. Ansley	Junior assistant engineer	1,800 per year	1,935 00		1,935 00
W. J. Bell	Junior assistant engineer	,320 per year	51 86	28 95	80 81
Wm. F Burke	Junior assistant engineer	,320 per year	50 74		50 74
W. J Burns	Junior assistant engineer	,800 per year	1,935 00		1,935 00
Charles L. Chapman	Junior asst. engineer (provisional)	,200 per year	93 55		93 55
A. B. Chappell	Junior assistant engineer	,680 per year	1,806 00		1,806 00
Edmund A. Close	Junior assistant engineer	,680 per year	480 63		480 63
J. F. Cullen	Junior assistant engineer	,200 per year	66 00		66 00
Thomas L. Curtin	Junior assistant engineer	,440 per year	521 13		521 13
B. S. Davenport	Junior assistant engineer	,560 per year	746 90		746 90
J. R. Eckhardt	Junior assistant engineer	,680 per year	479 61	7 65	487 26
J. Frank Egan	Junior assistant engineer	,320 per year	573 78		573 78
Fred C. Facer	Junior assistant engineer	1,800 per year	300 00		300 00
Joseph H. Friedman	Junior assistant engineer	1,320 per year	566 50		566 50
W. H. Ginnity	Junior assistant engineer	1,800 per year	1,935 00		1,935 00
Charles E. Heydt	Junior assistant engineer	1,680 per year	222 31		222 31
H. R. Horton	Junior assistant engineer	1,200 per year	35 36	31 46	66 82
Joseph W Howe	Junior assistant engineer	1,680 per year	1,682 50		1,682 50
Neil D. Hyde	Junior assistant engineer	1,560 per year	46 13		46 13
Alameth Kay	Junior asst. engineer (provisional)	1,560 per year	520 00		520 00
Edward J Kelley	Junior assistant engineer	1,440 per year	1,454 00		1,454 00
Fred G. Kimball	Junior assistant engineer	1,680 per year	1,667 50		1,667 50
Michael Kovar	Junior assistant engineer	1,320 per year	382 51		382 51
J. F. Larney	Junior assistant engineer	1,440 per year	208 87		208 87
L. John Long	Junior asst. engineer (provisional)	1,320 per year	1,439 90	226 91	1,666 81
Raymond M. Lynd	Junior asst. engineer (provisional)	1,200 per year	210 92		210 92
F. B. McLean	Junior assistant engineer	1,200 per year	858 00		858 00
Frank J. McMahon	Junior assistant engineer	1,300 per year	432 90		432 90
S. A. Miller	Junior assistant engineer	1,800 per year	1,811 50		1,811 50
W. R. Miller	Junior assistant engineer	1,440 per year	52 42		52 42
D. M. Miner	Junior assistant engineer	1,800 per year	574 84		574 84
J. E. Morrell	Junior assistant engineer	1,200 per year	60 68		60 68
C. V. O'Malley	Junior assistant engineer	1,800 per year	563 45		563 45
Jno. J Phalan	Junior assistant engineer	1,560 per year	1,222 39	2 05	1,224 44
O. J. Pierce	Junior assistant engineer	1,680 per year	1,129 27		1,129 27
W. W. Redfern	Junior assistant engineer	1,320 per year	75 00		75 00
Herbert S. Roberts	Junior assistant engineer	1,440 per year	344 96		344 96
H. A. Shafer	Junior assistant engineer	1,680 per year	1,035 10	5 23	1,040 33
D. T. Simpson	Junior assistant engineer	1,680 per year	280 00		280 00
J. A. Sloat	Junior assistant engineer	1,800 per year	777 90	51 01	828 91
Jacob Smertenko	Junior assistant engineer	1,320 per year	1,419 00		1,419 00
Tracy B. Smith	Junior assistant engineer	1,800 per year	1,550 88		1,550 88
Charles S. Sterling	Junior assistant engineer	1,560 per year	1,018 55		1,018 55
H. R. Topping	Junior assistant engineer	1,800 per year	165 76	19 38	185 14
Powell Wall	Junior assistant engineer	1,560 per year	20 43		20 43
C. M. Weinheimer	Junior asst. engineer (provisional)	1,200 per year	119 36		119 36
Edmond A. Weiss	Junior assistant engineer	1,320 per year	1,325 50		1,325 50
H. J Whitman	Junior assistant engineer	1,800 per year	1,935 00		1,935 00
Oswald R. Whyte	Junior assistant engineer	1,440 per year	130 88		130 88
J. F. Williamson	Junior asst. engineer (provisional)	1,200 per year	403 23		403 23
Alice E. Yale	Junior asst. engineer (provisional)	1,200 per year	153 23		153 23
W. J. Zabel	Junior assistant engineer	1,800 per year	1,214 01		1,214 01
Charles R. Zorsch	Junior assistant engineer	1,800 per year	1,935 00		1,935 00
Lynn B. Barrows	Engineering assistant	1,080 per year	665 64		665 64
Edgar M. Birdsall	Engineering assistant	840 per year	52 50		52 50
E. J. Bullis	Engineering assistant	1,020 per year	156 95		156 95
Walter R. Glock	Engineering assistant	900 per year	266 13		266 13
F. G. Hempel	Engineering assistant	1,080 per year	1,171 50		1,171 50
P. M. Howe	Engineering assistant	1,080 per year	126 41		126 41
Solomon Leibowitz	Engineering assistant	840 per year	77 00		77 00
W. F. Lysett	Engineering assistant	1,080 per year	264 00		264 00
Patrick J Murray	Engineering assistant (provisional)	840 per year	161 52		161 52
F. J. O'Connor	Engineering assistant	1,080 per year	606 77		606 77

*Includes additional compensation of 10 per cent allowed above base rate

Construction of Barge Canal — Erie Canal — (Concluded)

Chapter 147, Laws of 1903, and amendatory laws

NAME	Rank	Rate of compensation	*Services	Travel	Total
Joseph J. Raduciner....	Engineering assistant..........	$840 per year	$220 71	....	$220 71
Frank M. Sisson........	Engineering assistant..........	1,080 per year	137 32	....	137 32
Hugh J. Weir..........	Engineering assistant..........	1,080 per year	1,118 52	.....	1,118 52
George M. Harrer.....	Inspector of engineering works..	1,560 per year	150 97	....	150 97
James Sim............	Inspector of engineering works..	1,560 per year	1,677 00	....	1,677 00
W. A. Walter..........	Inspector of engineering works..	1,560 per year	1,677 00	....	1,677 00
Ernest F. Hamilton....	Boatman..............	3 00 per day	313 50	....	313 50
John Hano............	Boatman..............	3 00 per day	194 70	....	194 70
Samuel Kaiser........	Boatman..............	3 00 per day	171 60	....	171 60
Arthur Knapp........	Boatman..............	3 00 per day	184 80	....	184 80
C. Kumro............	Boatman..............	3 00 per day	636 90	....	636 90
J. H. McCabe........	Boatman..............	3 00 per day	650 10	....	650 10
M. McConnell........	Boatman..............	3 00 per day	960 30	....	960 30
John Riley..........	Boatman..............	3 00 per day	1,017 50	....	1,017 50
Richard Stanton......	Boatman..............	3 00 per day	1,036 20	....	1,036 20
Frank W. Wadley.....	Boatman..............	3 00 per day	1,151 70	....	1,151 70
William A. Brick......	Laborer..............	12 50 per day	635 25	....	635 25
William Brown.......	Laborer..............	50 per day	745 25	....	745 25
M. J. Connolly......	Laborer..............	50 per day	434 50	....	434 50
Sidney Z. Davidson....	Laborer..............	50 per day	38 50	....	38 50
George F. Doyle......	Laborer..............	50 per day	508 75	....	508 75
Edward H. Gleason....	Laborer..............	50 per day	178 75	....	178 75
Raymond J. Goldring..	Laborer..............	50 per day	242 00	....	242 00
Ralph Whitney Goosnell	Laborer..............	50 per day	148 50	....	148 50
Francis E. Green......	Laborer..............	50 per day	154 00	....	154 00
Smith Hulburt........	Laborer..............	50 per day	948 75	....	948 75
Henry J. Killian......	Laborer..............	50 per day	434 50	....	434 50
Lester Lavine........	Laborer..............	50 per day	880 00	....	880 00
Edward F. Murr......	Laborer..............	50 per day	860 75	....	860 75
Arthur W. Phillips....	Laborer..............	50 per day	165 00	....	165 00
Raymond J. Quandt....	Laborer..............	50 per day	756 25	....	756 25
Carlton F. Reule.....	Laborer..............	50 per day	591 25	....	591 25
Matthew Rigney......	Laborer..............	50 per day	437 25	....	437 25
Raymond J. Riley.....	Laborer..............	50 per day	500 50	....	500 50
George D. Rivers.....	Laborer..............	50 per day	860 75	....	860 75
W. H. Rundle........	Laborer..............	50 per day	357 50	....	357 50
Elmer R. Stoll........	Laborer..............	50 per day	46 75	....	46 75
Sydney W. Towe......	Laborer..............	50 per day	442 75	....	442 75
Lewis Van Allan......	Laborer..............	50 per day	863 50	....	863 50
E. R. Weed..........	Laborer..............	50 per day	893 75	....	893 75
John J. Nugent........	Chauffeur............	500 per year	1,612 50	$266 67	1,879 17
E. Quans............	Office assistant......	.020 per year	1,038 34	....	1,038 34
J. Horton Begy.......	Gage reader..........	120 per year	120 00	....	120 00
H. K. Compson.......	Gage reader..........	84 per year	84 00	....	84 00
C. H. Harrison.......	Gage reader..........	60 per year	60 00	....	60 00
Patrick J. Slavin......	Gage reader..........	60 per year	60 00	....	60 00
Homer Snell..........	Gage reader..........	72 per year	72 00	....	72 00
Carl Tuscher.........	Gage reader..........	60 per year	60 00	....	60 00
			$121,167 94	$4,153 07	$125,321 01

Incidental Expenses

Instruments and tools..	$101 72
Office rent...	3,350 50
Fuel and light..	152 16
Stationery and printing...	300 09
Postage..	381 43
Telephone and telegraph..	1,187 69
Miscellaneous..	7,686 38
	13,159 97

Total.. $138,480 98

* Includes additional compensation of 10 per cent allowed above base rate.

Construction of Barge Canal Terminals

Chapter 746, Laws of 1911, and amendatory laws

NAME	Rank	Rate of compensation	*Services	Travel	Total
B. E. Failing	Senior assistant engineer	$3,300 per year	$1,773 75	$76 35	$1,850 10
F. V. Searls	Estimate clerk	1,920 per year	389 51	5 77	395 28
W. D. Gartland	Stenographer	1,200 per year	152 58		152 58
Mary MacArthur	Stenographer	1,200 per year	770 00		770 00
M. Agnes Maloney	Stenographer	900 per year	295 40		295 40
Elias H. Anderson	Assistant engineer	2,580 per year	2,169 54	62 29	2,231 83
C. E. Elmendorf	Assistant engineer	2,580 per year	748 04		748 04
F. W. Madigan	Assistant engineer	2,580 per year	116 22		116 22
Frank T. Marsh	Assistant engineer	2,340 per year		5 82	5 82
H. N. Metzger	Assistant engineer	2,340 per year	374 56	124 67	499 23
Lester P. Slade	Assistant engineer	2,160 per year	95 81		95 81
Elwin G. Speyer	Assistant engineer	2,580 per year	1,847 98	135 00	1,978 98
J. S. Summers	Assistant engineer	2,580 per year	466 14	1 33	2,484 47
F. J. Wilbur	Assistant engineer	2,580 per year	2,554 80		,554 80
C. J. Bean	Junior assistant engineer	1,800 per year	1,911 43	2 15	,913 58
C. B. Bennett	Junior assistant engineer (provisional)	1,200 per year	116 13		116 13
Byron T. Bisgood	Junior assistant engineer (provisional)	1,560 per year	65 00		65 00
Thomas L. Curtin	Junior assistant engineer	1,320 per year	230 39		230 39
Walter G. Dubey	Junior assistant engineer	1,560 per year	1,017 07		1,017 07
Charles E. Heydt	Junior assistant engineer	1,560 per year	212 19		212 19
Neil D. Hyde	Junior assistant engineer	1,560 per year	286 00		286 00
F. B. McLean	Junior assistant engineer	1,200 per year	110 00		110 00
W. R. Miller	Junior assistant engineer	1,560 per year	76 27		76 37
C. V. O'Malley	Junior assistant engineer	1,800 per year	58 55		58 55
Jno. J. Phalan	Junior assistant engineer	1,440 per year	161 31		161 31
O. J. Pierce	Junior assistant engineer	1,560 per year	538 23		538 23
W. W. Redfern	Junior assistant engineer	1,200 per year	443 55		443 55
M. B. Severance	Junior assistant engineer	1,560 per year	419 35		419 35
J. A. Sloat	Junior assistant engineer	1,800 per year	88 00		88 00
Tracy B. Smith	Junior assistant engineer	1,800 per year	378 79		378 79
I. L. Stalker	Junior assistant engineer	1,560 per year	1,035 04		1,035 04
C. S. Sterling	Junior assistant engineer	1,440 per year	314 51		314 51
H. R. Topping	Junior assistant engineer	1,800 per year	141 23	97 86	239 09
Powell Wall	Junior assistant engineer	1,680 per year	96 61		696 61
Oswald R. Whyte	Junior assistant engineer	1,320 per year	84 28		584 28
George M. Harrer	Inspector of engineering works	1,560 per year	1 26 03		1,526 06
Samuel Kaiser	Boatman	3 00 per day	57 70		557 70
Arthur Knapp	Boatman	3 00 per day	23 40		323 40
J. H. McCabe	Boatman	3 00 per day	36 30		36 30
David R. Petrikin	Boatman	3 00 per day	1,032 90		1 032 90
George F. Doyle	Laborer	2 50 per day	258 50		258 50
W. H. Rundle	Laborer	2 50 per day	503 25		503 25
			$24,876 24	$524 24	$25,400 48

Incidental Expenses

Office rent	$1,007 00
Fuel and light	32 54
Stationery and printing	2 90
Postage	33 19
Telephone and telegraph	107 25
Miscellaneous	397 58
	1,580 46

Total $26,980 94

*Includes the additional compensation of ten per cent allowed above the base rate.

Chadakoin River Improvement

Chapter 758, Laws of 1913; chapter 728, Laws of 1915; chapter 181, Laws of 1917; chapter 644, part 5, Laws of 1919

NAME	Rank	Rate of compensation	*Services	Travel	Total
H. N. Metzger........	Assistant engineer..............	$2,340 per year	$173 76	$45 30	$219 06
W. W. Redfern.......	Junior assistant engineer.......	1,200 per year	7 21	....	7 21
Powell Wall........	Junior assistant engineer.......	1,560 per year	64 58	....	64 58
Oswald R. Whyte......	Junior assistant engineer.......	1,320 per year	141 95	8 45	150 40
Samuel Kaiser........	Boatman...................	3 00 per day	3 30	....	3 30
Horace S Butts.......	Gage reader..............	120 per year	110 00	....	110 00
			$500 80	$53 75	$554 55

Incidental Expenses

Miscellaneous..		50
Total..		$555 05

Ellicott Creek Improvement

Chapter 624, Laws of 1913; chapter 728, Laws of 1915; chapters 181 and 760, Laws of 1917

NAME	Rank	Rate of compensation	*Services	Travel	Total
R. W. Cady..........	Assistant engineer..............	$2,580 per year	$696 73	$3 05	$696 78
D. T. Simpson........	Junior assistant engineer.......	1,680 per year	319 67	....	319 67
C. M. Weinheimer.....	Junior assistant engineer (provisional).................	1,200 per year	100 00	....	100 00
Lynn H. Barrows......	Engineering assistant...........	1,080 per year	438 69	....	438 69
C. Kumro...........	Boatman...., ...	3 00 per day	402 60	....	402 60
W. A. Brick..........	Laborer...................	2 50 per day	162 25	....	162 25
			$2,116 94	$3 05	$2,119 99

Incidental Expenses

Office rent...	$42 00	
Stationery and printing...	1 04	
Postage..	2 97	
Telephone and telegraph...	11 10	
Miscellaneous..	18 69	
		75 80
Total..		$2,195 79

* Includes the additional compensation of ten per cent allowed above the base rate.

Hertel Avenue Bridge, Buffalo
Chapter 761, Laws of 1917

NAME	Rank	Rate of compensation	*Services	Travel	Total
Elias H. Anderson....	Assistant engineer............	$2,580 per year	$104 26	$21 20	$125 46
Gordon Edson.........	Assistant engineer...........	2,580 per year	68 66	....	68 66
Lester P. Slade........	Assistant engineer...........	,100 per year	57 48	....	57 48
F. J. Wilbur..........	Assistant engineer...........	,580 per year	61 03	....	61 03
Frank V. Searls.......	Estimate clerk...............	,920 per year	51 10	....	51 10
C. J. Bean..........	Junior assistant engineer........	,800 per year	23 57	1 30	24 87
Walter G. Dubey.....	Junior assistant engineer........	,440 per year	491 93	....	491 93
Neil D. Hyde........	Junior assistant engineer........	,560 per year	143 00	....	143 00
I. L. Stalker.........	Junior assistant engineer........	,560 per year	470 12	....	470 12
Oswald R. Whyte....	Junior assistant engineer........	,200 per year	14 20	....	14 20
E. Quam............	Office assistant...........	1,020 per year	15 08	....	15 08
			$1,500 43	$22 50	$1,522 93

Incidental Expenses		
Postage..	$0 49	
Telephone and telegraph.......................	40	
Miscellaneous..................................	101 96	
		102 85
Total.........................		$1,625 78

Eighteen-Mile Creek Culvert, Lockport
Chapters 181 and 626, Laws of 1917

ITEM	Total
Incidental Expenses	
Stationery and printing..	$55 80

Griffin Creek Improvement, Cuba
Chapter 565, Laws of 1918

NAME	Rank	Rate of compensation	*Services	Travel	Total
C. E. Elmendorf......	Assistant engineer............	$2,580 per year	$212 06	$117 31	$329 37
Jno. J Phalan	Junior assistant engineer........	1,440 per year	35 19	....	35 19
Hugh J Weir........	Engineering assistant...........	1,020 per year	6 23	....	6 23
E. F. Murr..........	Laborer...................	2 50 per day	8 25	....	8 25
			$261 73	$117 31	$379 04

Incidental Expenses		
Livery..	$5 00	
Miscellaneous..................................	13 36	
		18 36
Total.........................		$397 40

* Includes the additional compensation of ten per cent allowed above base rate.

Blue Line Surveys
Chapter 151, Laws of 1913

NAME	Rank	Rate of compensation	*Services	Travel	Total
L. S. Hulburd	Senior assistant engineer	$3,300 per year	$1,697 91	$57 58	$1,755 49
Mary MacArthur	Stenographer	1,200 per year	220 00	.....	220 00
O. L. Burdett	Assistant engineer	2,340 per year	471 90	.....	471 90
C. E. Elmendorf	Assistant engineer	2,580 per year	1,166 97	.....	1,166 97
L. G. Fisher	Assistant engineer	2,340 per year	826 38	.....	826 38
Frank T. Marsh	Assistant engineer	2,340 per year	242 18	.....	242 18
F. J. Wilbur	Assistant engineer	2,580 per year	157 67	.....	157 67
R. W. Anderson	Junior assistant engineer	1,440 per year	25 01	.....	25 01
Wm. F. Burke	Junior assistant engineer	1,320 per year	80 67	.....	80 67
Michael Kovar	Junior assistant engineer	1,440 per year	165 94	..	165 94
W. R. Miller	Junior assistant engineer	1,680 per year	603 37	.....	603 37
C. V. O'Malley	Junior assistant engineer	1,800 per year	779 94	.	779 94
J. A. Sloat	Junior assistant engineer	1,800 per year	556 03	19 75	575 78
I. L. Stalker	Junior assistant engineer	1,680 per year	148 50	.....	148 50
H. R. Topping	Junior assistant engineer	1,800 per year	725 59	74 83	800 42
E. J. Bullis	Engineering assistant	1,080 per year	412 53	.....	412 53
Edward J. Moran	Engineering assistant	960 per year	85 25	.....	85 25
E. Quans	Office assistant	1,020 per year	9 36	.....	9 36
J. H. McCabe	Boatman	3 00 per day	323 40	.....	323 40
M. McConnell	Boatman	3 00 per day	141 90	.....	141 90
Freeman S. Barclay	Laborer	2 50 per day	41 25	.....	41 25
R. J. Quandt	Laborer	2 50 per day	77 00	.....	77 00
Carlton F. Reule	Laborer	2 50 per day	41 25	.	41 25
			$9,000 00	$152 16	$9,152 16

Incidental Expenses

Stationery and printing	$20 82
Livery	143 50
Postage	39 46
Office rent	774 50
Telephone and telegraph	60 96
Miscellaneous	308 61
	1,347 84
Total	$10,500 00

Surveys for State Court of Claims
Chapter 151, Laws of 1913

NAME	Rank	Rate of compensation	*Services	Travel	Total
Elias H. Anderson	Assistant engineer	$2,580 per year	$16 89	$28 39	$45
W. W. Brown	Assistant engineer	2,160 per year	92 90	6 18	99 1
Gordon Edson	Assistant engineer	2,580 per year	.....	1 72	1 7
C. E. Elmendorf	Assistant engineer	2,580 per year	28 67	3 00	31 67
H. N. Metzger	Assistant engineer	2,340 per year	473 35	81 07	554 42
Lester P. Slade	Assistant engineer	2,160 per year	12 77	.....	12 77
Elwin G. Speyer	Assistant engineer	2,340 per year	678 52	63 97	742 49
J. S. Summers	Assistant engineer	2,580 per year	22 88	2 50	25 38
L. John Long	Junior assistant engineer	1,320 per year	12 10	2 95	15 05
F. B. McLean	Junior assistant engineer	1,200 per year	3 55	.....	3 55
Tracy B. Smith	Junior assistant engineer	1,800 per year	5 33	6 11	11 44
Powell Wall	Junior assistant engineer	1,560 per year	35 75	.....	35 75
Oswald R. Whyte	Junior assistant engineer	1,320 per year	80 99	.....	80 99
Samuel Kaiser	Boatman	3 00 per day	36 30	.....	36 30
			$1,500 00	$195 89	$1,695 89

Incidental Expenses

Postage	$44 77
Office rent	370 00
Telephone and telegraph	32 05
Miscellaneous	7 29
	454 11
Total	$2,150 00

* Includes the additional compensation of ten per cent allowed above base rate.

Eighteen-Mile Creek Survey, Niagara County

Chapter 425, Laws of 1918

NAME	Rank	Rate of compensation	*Services	Travel	Total
H. N. Metzger	Assistant engineer	$2,340 per year	$556 27	$373 75	$930 02
Byron T. Blagood	Junior assistant engineer (provisional)	1,560 per year	91 00	....	91 00
F. B. McLean	Junior assistant engineer	1,200 per year	348 45	....	348 45
L. L. Stalker	Junior assistant engineer	1,560 per year	13 34	....	13 34
Oswald R. Whyte	Junior assistant engineer	1,320 per year	310 34	....	310 34
Alice E. Yale	Junior assistant engineer (provisional)	1,200 per year	61 29	96 60	157 89
Lynn H. Barrows	Engineering assistant	1,080 per year	83 67	....	83 67
Samuel Kaiser	Boatman	3 00 per day	211 20	....	211 20
Wm. A. Brick	Laborer	2 50 per day	13 75	....	13 75
			$1,690 31	$470 35	$2,160 66

Incidental Expenses

Livery	$132 25	
Postage	1 34	
Miscellaneous	39 49	
		173 08
Total		$2,333 74

SUMMARY

The foregoing tables are summarized as follows:

Ordinary Repairs to Canal

1. Erie canal, chapter 151, Laws of 1918 .... $10,000 00

Construction of Barge Canal

2. Erie canal, chapter 147, Laws of 1903, and amendatory laws 138,480 98

Construction of Barge Canal Terminals

3. Erie canal, chapter 746, Laws of 1911, and amendatory laws 26,980 94

Special Work

4. Chadakoin river improvement, chapter 758, Laws of 1913; chapter 728, Laws of 1915; chapter 181, Laws of 1917; chapter 644, part 5, Laws of 1919 ....   555 05
5. Ellicott creek improvement, chapter 624, Laws of 1913; chapter 728, Laws of 1915; chapters 181 and 760, Laws of 1917 .... 2,195 79
6. Hertel avenue bridge, Buffalo, chapter 761, Laws of 1917 .... 1,625 78
7. Eighteen-Mile creek culvert, Lockport, chapters 181 and 626, Laws of 1917 ..   55 80
8. Griffin creek improvement, Cuba, chapter 565, Laws of 1918 .... 397 40

Special Surveys

9. Blue line surveys, chapter 151, Laws of 1918 .... 10,500 00
10. Surveys for State Court of Claims, chapter 151, Laws of 1918 .... 2,150 00
11. Eighteen-Mile creek survey, Niagara county, chapter 425, Laws of 1918 ..... 2,333 74

     Total $195,275 48

* Includes the additional compensation of ten per cent allowed above the base rate.

REPORT OF TESTS

REPORT OF THE LAND BUREAU

REPORT OF TESTS

TESTING LABORATORY — GEOLOGICAL HALL

ALBANY, N. Y., *July* 1, 1919.

HON. FRANK M. WILLIAMS, *State Engineer and Surveyor:*

Sir.— I have the honor to submit the following report of the work of the testing laboratory of your Department for the fiscal year ended June 30, 1919.

The work of the laboratory during the past year has been more extensive than the year before, for, not only did the volume of work increase, but the variety of the work also increased. As heretofore, the testing of cement and other concrete materials forms the bulk of the testing, but the ratio of cement testing to other testing is gradually decreasing. The tests and analyses of the wide variety of other construction materials makes the laboratory of special and an increasing value for its effect in all phases of constructions. Opportunities for tests and research on the effect of time and weather on several of these materials secure for the laboratory a knowledge of value for use by various State departments.

CEMENT TESTS

The work of testing the cement proposed for use on Barge canal work and on some other State works was almost 40 per cent greater this year than last. During the year there have been submitted 3,656 samples of cement, representing 218,533 barrels of cement, of which 96 per cent was tested for the State Engineer's Department and 4 per cent for other State works. It has been found that for promptness of inspection and delivery as well as economy it was advisable to permit shipments frequently to the State Architect's department and to other State work from some bin of cement which had been tested and accepted for use on Barge canal work. Because of this practice considerable more cement was shipped under our inspection to these other works from " Barge canal " bins than the 4 per cent noted above.

The inspection of cement at the mills has long been a large part of the work of this bureau. Such inspection permits the taking of a smaller proportion of samples to the number of barrels represented.   In addition to the saving of time and the number of necessary samples, mill inspection prevents the delivery of any cement upon the work except that which has been tested and accepted.   Notwithstanding these advantages there continues to be a considerable amount of cement sampled on the work after delivery and then tested, for some contractors prefer to have it done that way and the specifications permit the contractor to choose. They frequently choose the latter way in the effort to save the per barrel charge made by cement companies for storage in the bins put under the seal of " bin-tested " cement.

Each sample of cement submitted, mixed in the proportion of one part cement to three parts standard quartz sand, was tested for tensile strength at 7-day and 28-day periods.   In addition to the tests for tensile strength, each lot of samples was given tests for fineness of grinding, for initial and hard sets, for specific gravity and for soundness, by means of the steam tests, the normal-water test and the normal-air test.   Frequently the cements were completely analyzed and are specially checked for sulphuric anhydride ($S O_2$) and magnesia ($Mg O$).

The methods used in making the tests and analyses of cement are those adopted as standard by the American Society for Testing Materials.   A slight variation, however, is that, instead of using a blended sample for tests for tensile strength, we still use our own method of testing each sample separately for tensile strength. This method has long proven very satisfactory; in fact, by means of it much poor cement has been discovered which would have stood the tests had all of the samples of a lot been blended and then tested.   It has been a special help in securing from the mills a cement that is uniform in quality.   This method, however, makes necessary a larger equipment and a more complete system of operation than is necessary under the common method of testing the blended sample.   This large and splendid equipment we have in our laboratory.   The effort has constantly been made to maintain this complete laboratory with as little expenses as possible, and this laboratory has earned a wide reputation for the accuracy of its results and the efficiency of its inspections.

The specifications of this Department for cement are now those adopted by the American Society for Testing Materials. While formerly this Department used a crushed quartz sand in making its tests for tensile strength, this laboratory now uses only the standard Ottawa sand. This it can do now, since the contracts let with the specification calling for tests with the crushed quartz sand are all completed and all the active and new contracts call for the use of Ottawa sand.

All results obtained at the end of the 7-day tests of the samples of cement proposed for use on Barge canal work are reported to Mr. F. P. Williams, Special Deputy State Engineer, and, if then thought best, are held for the 28-day tests, the lots being accepted or rejected by him as the results show that the cement passes or fails to pass the requirements of the tests. The reports of all tests of cement for all other Department work (except Barge canal) are submitted to the Deputy State Engineer, Mr. R. G. Finch.

Because this laboratory — through its director — is in such close relationship to the Committee on Cement of the American Society for Testing Materials it has also been making a thorough study of the proposition to substitute compression tests of cement for the present tension tests and it is securing data which will help toward a wise decision on this question.

Of the cement tested and proposed for use all was Portland cement. Nineteen brands of cement were tested and of these 5 were manufactured in New York, 12 in Pennsylvania, 1 in Ohio and 1 in New Jersey.

The method of inspection of cement at the various mills is as follows: When there is to be enough cement to warrant doing so, an inspector is sent to the cement mill to sample cement and inspect shipments. The inspector takes samples from the various parts of the bin or from the conveyor as the cement is being carried to the bin, and each sample is tested in the same way as are the samples taken from cement delivered on the work. The endeavor is to obtain from the sampling and the testing of these samples the "run of the product." As soon as the samples are taken, the inspector places the bin of cement under the seal of this Department and the bin is so sealed that no cement can be added to or taken from it without detection. When the results of the tests have been secured, the reports are made in the usual way, and then, if the cement is accepted, the bin of cement is assigned to the contract which may have placed an order for the cement. When the contractor needs cement, the inspector at the mill breaks the seal on the bin, inspects the loading of the car or cars, seals these with the Department seal and then reseals the bin of cement. A notice of shipment is forwarded to the laboratory, is examined and approved, if correct, and sent to the senior assistant engineer in charge of the contract to which the cement has been assigned. When the car or cars arrive on the work, the seal of the Department must be broken by the senior assistant engineer in charge or his representative, otherwise the lot of cement must be sampled and tested in the usual way.

SAND TESTS

The thorough examination and tests of the sands and gravels proposed for use on work in the various departments have been continued and the importance of such tests has frequently been fully demonstrated. It has been found that almost all of the available sand and gravel banks along the line of the canal sytsem have been sampled and tested, and with these it is now only necessary to make occasional inspections and tests to ascertain whether or not the quality of the materials from these banks is equal to the samples accepted. Dredging the canal channel has frequently opened a section in which sand and gravel has been of such quality that it has been proposed for use. Spoils-banks made of this material have been examined and tested for the quality of the sand. Many other banks throughout the State, but not along the canal system, have also been tested. The results of the tests of these materials have also been found to be of value by other State departments.

The tests made are as follows: The sands are examined under the microscope for those elements that give the sand its characteristics. The other tests are for voids, loam and silt, fineness or grading, and strength—both tensile and compressive—with cement. The latter are made from the sand in its natural condition and also washed; and the cement used is a "standard" cement, made by mixing together in the laboratory several brands of cement which have given results nearly alike in the regular tests. All tests for strength cover at least 28 days, but many long-time tests are being carried. Considerable attention has been given to the methods used in making the tests and it is believed that the most accurate methods are being used.

The testing of sand includes also the testing and examination of the gravel in the sand and also the testing of substitutes for sand and gravel, such as screenings, iron-ore tailings and slag.

CONCRETE TESTS

Along with the testing of the cement and fine aggregates, tests have been carried along on coarse aggregates, such as various kinds of crushed rocks, gravels, slag, etc., by means of compression tests on concrete made up of these materials. Tests for the effects of varying the proportions, the consistencies of the mixes, the methods of molding and capping and of storing the test-pieces have been made. Also tests have been made of samples of concrete made up as it was being placed in the structures and sent in from the various works.

TESTS OF OTHER MATERIALS

Besides those already reported, there have been made a large variety of tests and analyses of other construction and building materials. Among the materials thus examined were stones, artificial building stone, mortars and colorings, wooden paving blocks, granite and sandstone paving blocks, paving bricks, face bricks, chimney brick, both paving and roofing bituminous materials, hollow tile, galvanized conduits and fixtures, waterproofings of various types, wood preservatives, paints, varnishes and puttys. Research work has been continued on laitance and on the efflorescence and incrustations on concrete, and microscopic examinations and analyses in the laboratory on these materials have also been continued.

FIELD INSPECTIONS

In addition to directing the work of the laboratory and the mill inspections, the undersigned has made field inspections of the concrete and concrete materials being used on many of the Barge canal contracts where concrete was being placed. Particular attention has been given to the sources of supply of the gravels, sand and stone being used in the concrete. A more definite knowledge is thus gained than is possible through a laboratory sample alone, but with both tests and field inspection absolute knowledge of the materials is gained. Inspection of its actual use is also a help in considering the points of merit or demerit in the material. Inspections of concrete that has been in place for some time have been made in order to study the condition and the wear of the concrete, for the purpose of securing information on the various theories that have been advanced from time to time on the changes that may take place in concrete. Some interesting and profitable studies have thus been made. The construction of concrete barges for proposed use on the Barge canal have also been watched and inspected during the various stages of their construction.

The development of the specifications for and the methods of tests of the materials of construction is a natural sequence of the knowledge secured in the analyses, tests and inspections, so this has frequently been a feature of the work we have been called upon to do.

Respectfully submitted,

RUSSELL S. GREENMAN,

Senior Assistant Engineer, in charge of Tests.

REPORT OF THE LAND BUREAU

STATE OF NEW YORK

DEPARTMENT OF THE STATE ENGINEER AND SURVEYOR

LAND BUREAU

ALBANY, N. Y., *July* 1, 1919.

Hon. FRANK M. WILLIAMS, *State Engineer and Surveyor:*

Sir.— Herewith I submit a report of the work of the Land Bureau for the fiscal year ended June 30, 1919.

The sale of State land that is ordered sold by the Commissioners of the Land Office is conducted by this bureau. Ten public auctions were held and the sum of $7,291.55 realized therefrom. A detailed statement of the sales is appended. The average sale for the previous ten years is $11,835.00.

Maps of all grants of land under water made by the Commissioners of the Land Office and the Legislature are on file in this bureau and new grants are added when made by the Land Board.

The early records, maps and field notes filed in this bureau are being constantly examined by the public and are of great and increasing value.

Twelve modern atlases of various counties of the State have been added to the library during the year and the need of additional room is felt.

Respectfully submitted,

MERRITT PECKHAM, JR.

Land Clerk.

TABLE OF SALES CONDUCTED BY THE LAND BUREAU DURING THE FISCAL YEAR ENDED JUNE 30, 1919

Date of sale	Purchaser	Location	Lot	Acres	Tax or mortgage	Price
1918						
July 11	Jaspare Sciuto.......	Lockport........	1 city lot	......	Tax	$133 00
July 16	Tunis A. Swick.......	Tompkins county.	....	4.037	Tax	277 00
Aug. 6	Louie Smith.........	Chautauqua county........	....	66	Mortgage	802 00
Aug. 9	Wesley W. Sternberg..	Madison county..	....	75	Mortgage	825 00
Oct. 1	Standard Oil Company	Syracuse........	1 city lot	......	Original	690 00
Nov. 13	Carrie Askanasy.....	Kings county....	2 city lots	......	Tax	48 50
1919						
Jan. 7	Leslie W. Paine......	Ontario county. .	....	156	Mortgage	1,500 00
Mar. 5	Cathrine Noonan.....	Kings county....	3 city lots	......	Tax	76 05
April 9	City of Syracuse.....	Syracuse........	1 city lot	......	Original	2,707 00
May 29	Crucible Steel Co.....	Syracuse........	1 city lot	......	Canal	233 00
	Total..........	....	....	......	....	$7,291 55

REPORT

ON

EXAMINATION AND SURVEY OF EIGHTEEN-MILE CREEK WITH ESTIMATES OF COST OF PROPOSED IMPROVEMENTS

[225]

REPORT ON WORK DONE UNDER CHAPTER 425 OF THE LAWS OF 1918, AUTHORIZING THE MAKING OF A SURVEY OF EIGHTEEN-MILE CREEK IN THE COUNTY OF NIAGARA AND MAKING AN APPROPRIATION THEREFOR

By the above named act the State Engineer and Surveyor was authorized " to make such examination and survey of Eighteen-Mile creek north of the line of the Erie canal in the county of Niagara as may be necessary to determine what improvements in the way of channel deepening or straightening of banks may be required to enable the said creek to carry flood waters and overflow from the Erie canal without damage to private property abutting on said creek, and to prepare an estimate of the cost of all such work and improvement."

Under the direction of Mr. F. P. Williams, Division Engineer of the Western Division, Mr. Byron E. Failing, Senior Assistant Engineer, had charge of the making of the necessary surveys, maps and estimates.

Information on file in the Department of the State Engineer was supplemented by field surveys.

Description of Stream

Eighteen-Mile creek rises about two miles south of the city of Lockport and flows northerly through that city, emptying into Lake Ontario at Olcott. It has a drainage area of 85 square miles. The stream passes under the Barge canal about half a mile east of the Lockport locks and for the next 1½ miles it passes through a gorge with a total fall of about 150 feet. From this point to Lake Ontario it has a uniform slope with a fall of 119 feet in about 14 miles as measured along its course.

The drainage area of the creek south of the canal is too small to furnish sufficient water for a practicable water-power development. Water from the canal, however, is wasted into it in sufficient volume to make the development of hydraulic power profitable, and several mills and factories have been located on the stream to utilize such power.

WATER-POWER SITUATION AT LOCKPORT

Between the upper and lower pools of the Barge canal at Lockport there is a fall of about 50 feet. For feeding the canal east of Lockport it is necessary to by-pass a considerable amount of water around the locks. The head being large, the value of the water for power purposes was quickly realized, early in the history of the original canal, and the right to use that water was leased by the State under what is known as the "Kennedy and Hatch lease" in 1826. The terms of this lease are are follows:

Kennedy and Hatch Lease

This indenture made this 25th day of January in the year 1826, between the Canal Commissioners of the State of New York of the first part, and Richard Kennedy of the Town of Lockport and Junius H. Hatch of the City of New York of the second part witnesseth:

That the said canal commissioners in pursuance of the provisions of the act entitled "An Act concerning the Erie and Champlain Canal" and passed April 20th, 1825, and in consideration of the sum hereinafter covenanted to be paid by said Kennedy and Hatch have granted, leased and to farm let, and by these presents do grant, lease and to farm let to said Kennedy and Hatch, their heirs, executors, administrators, and assigns all the surplus waters which without injury to navigation or the security of the canal may be spared from the canal at the head of the locks in the Village of Lockport, to be taken and drawn from the canal at such place and in such manner and discharged into the lower level at such places and in such manner as said canal commissioners shall from time to time deem most advisable for the security of the canal and for the convenience of navigation thereof.

And the said Kennedy and Hatch hereby jointly and severally covenant and engage to pay to the commissioners of the canal fund yearly and every year hereafter on the 1st of January in each year the sum of Two hundred dollars and for the eventual payment thereof they hereby bind themselves, their heirs, executors, administrators and assigns.

It is hereby expressly understood and agreed that the canal

commissioners reserve to themselves and to the legislature the right to limit, control, or wholly resume said waters and all the rights granted by this lease whenever in the opinion of said canal commissioners or of the legislature the safety of the canal and its appendages or the necessary supply of water for the navigation of the canal shall render such limitation, control and resumption necessary.

It is further agreed that if at any time the rent hereby reserved shall remain unpaid for one year after the same shall become due that this lease shall be forfeited to the state and the said commissioners may thereupon relet the said surplus waters to any other person in like manner as if no lease thereof had been executed.

In Witness Whereof the parties of these presents have hereunto set their hands and seals the day and year first above written.

R. KENNEDY [SEAL]

JUNIUS H. HATCH [SEAL]

SAMUEL YOUNG

HENRY SEYMOUR [SEAL]

W. C. BOUCK

Canal Commissioners.

The Hydraulic Race Company is the successor in title to this lease, but it does not claim to own all of the surplus waters. Disputes having arisen between the lessees and other users of surplus water, litigation ensued and on August 27, 1851, a decree was handed down by the Supreme Court in Buffalo fixing the title in the three other parties to a portion of the surplus water in the power canals and tunnels at Lockport. These rights have changed hands several times and at present there are two mills claiming rights which are independent of those of the Hydraulic Race Company. These are the Thompson Milling Company with 96 horse-power and the Franklin Mills Company with the same amount of power. The Hydraulic Race Company does not develop power itself, but leases power rights to the following concerns: Lockport Light, Heat & Power Co., Grigg Bros. Flour Mill, United Paper Board Co., Trevor Manufacturing

Co., Western Block Co., and Thompson Milling Co., which develop in all about 4,000 horse-power and use approximately 900 cubic feet of water per second.

Use and Diversion of Canal Water at Lockport

It is estimated that the maximum amount that will be required for the Barge canal below Lockport will be 1,237 second-feet. Of this, 230 second-feet will be required for lockage with a canal traffic of 10,000,000 tons annually and 400 second-feet with a maximum canal traffic, leaving under these conditions about 900 second-feet as surplus water available for power between the upper and lower levels.

It is apparent, therefore, that the surplus water passing the locks for navigation purposes is practically all used for power at the present time. In addition to the water needed for navigation purposes a diversion of 500 second-feet from the Niagara river by way of Tonawanda creek and the canal was granted by the Secretary of War on August 16, 1907, to the Lockport Hydraulic Company, or as it is now known, the Hydraulic Race Company.

On November 25, 1913, a revocable permit was granted by the State to the Lockport and Newfane Mill Owners Association, Inc., in which the State agreed to discharge the 500 second-feet granted by the Secretary of War into Eighteen-Mile creek, or as much thereof as the creek could carry without damage being done. For this service the State was to receive $7,500 per annum. If for any reason this amount of water could not be furnished, then a pro-rata deduction was to be made, based upon the quantity of water delivered. The Lockport and Newfane Mill Owners Association, Inc., is a corporation composed of about all of the mill owners on Eighteen-Mile creek, together with the Hydraulic Race Company.

EIGHTEEN-MILE CREEK
Capacity of Present Channel

Under Mr. Failing's direction the capacity of Eighteen-Mile creek was found by passing different known quantities down the stream and observing the results. Current-meter measurements were taken at several places to ascertain the velocity. The stream was cross-sectioned and gage stakes placed about 1,000 feet apart. It was thus ascertained that the channel of the creek would carry 375 second-feet without overflowing its banks. The height of water with 375 second-feet passing was observed and from this information a hydraulic grade line for the water surface was computed for flows of 500 second-feet and 1,000 second-feet with the stream channel in its present condition. Computations were also made to determine the amount of excavation necessary to carry these discharges at the same water-surface elevation as that given by the flow of 375 second-feet. The results of these computations are given in the attached tables, Nos. 1 and 2.

In Tables Nos. 1 and 2 the first column shows the number of the range at which the readings were taken; column No. 2, the distances in feet between ranges; column No. 3, the cubic yards of excavation to enlarge the channel so that the water-surface will not be raised above that for the discharge of 375 second-feet; column No. 4, the estimated excavation in cubic yards per linear foot of stream; column No. 5, the area needed for spoil banks; column No. 6, description of work to be done on the present channel, and column No. 7, the area of land in acres that would be flooded if no excavation is done in the present channel and additional water is allowed to flow over and flood the adjacent banks.

From Olcott to a point about two miles above Collins dam the creek has a sufficient capacity to carry 500 second-feet without doing damage. From this point to the Jackson street bridge it will be necessary to either appropriate the lands flooded or enlarge the channel. If the channel is enlarged, it should be done by widening between ranges 25 and 14, as in that section there is rock in or near the bed of the stream. From range 14, near the Turnpike road, to the Jackson road bridge, at the mouth of the gorge at the Lockport city line, the stream should be deepened instead of widened, as there is no rock near the surface and the deeper channel gives a better section for discharge and at less cost.

The removal of a few bends will not materially increase the capacity of the stream, for it is shallow and has a very rapid flow for its full distance. The straightening of a bend in such a stream improves the condition only in that immediate location and the cost will greatly exceed that of widening or deepening.

From Jackson's bridge to the Barge canal, a distance of a little over one mile, there are eight water-powers and over this section there is no chance of the creek causing damage unless one or more of the dams are raised.

Water-Power Available

Table No. 3 is a tabulation of the developed and undeveloped power on Eighteen-Mile creek, with a flow of 375 second-feet. This table shows that at present with a flow of 375 second-feet there is developed on Eighteen-Mile creek a total of 2,070 horse-power and there is available but undeveloped a total of 7,090 horse-power, making a total of developed and undeveloped power of 9,160 horse-power. In other words, 77 per cent of the available power on Eighteen-Mile creek with a flow of 375 second-feet is undeveloped and not used.

Table No. 4 shows similar data for the flow of 500 second-feet. That is, the development at present is 2,070 horse-power and the undeveloped power 10,150 horse-power or a total possible development of 12,220 horse-power. It should be borne in mind that the amounts of undeveloped power are based upon the total possible fall at the various sites and might not be fully realized in actual construction.

By personal interviews with the various power owners or their managers Mr. Failing obtained the following data regarding the present condition of the various plants:

Present Condition of Plants

50-FOOT HEAD AT LOCKS.

Lockport Light, Heat & Power Co.   Mr. Ferry, Chief Operator.

Has two water wheels; old one 500 hp., new one 1,500 hp. Obtains water from the Hydraulic Race Co.

Old wheel uses about 149 sec.-ft. and new one about 325 sec.-ft. (information from C. E. Dickenson).

Grigg Brothers Flour Mill.

Has one 75-hp. wheel, four or five years old, and uses about 22 sec.-ft., or 68 hp.

Has a perpetual lease for 35 hp. (of water) and buys an additional 33 hp. from the Hydraulic Race Co.

Thompson Milling Co.   Mr. Whitbeck, Manager.

Has one 650-hp. wheel, Holyoke Machine Co., Maker, installed about two years ago.   Is using the following power:

Its own	98 horse-power.
Perpetual lease	78 horse-power.
Lease from Hydraulic Race Co..........	294 horse-power.
Total	470 horse-power.

Trevor Manufacturing Co.

One 21-hp. wheel, using 20 hp., for which it has a perpetual lease.

Niagara Preserving Co.   Now owned by Western Block Co.

Has one new wheel, rated at 135 hp. capacity, built by S. Morgan Smith Co.   Not now operating.

Western Block Co.

Has one old 85-hp. wheel, using 75 hp. for which it pays rent to the Hydraulic Race Co.

(See also Niagara Preserving Co.)

Niagara Emery Co.   Owned by the Franklin Mills Co. of Batavia, N. Y.

Has one old 120-hp. wheel, and owns 96 hp., for which it pays no rent.   Is now using about 75 hp.

United Paper Board Co.

In its pulp mill at the canal, has two old wheels, rebuilt, at 1,200 = 2,400 hp. capacity.   Is now using 250 sec.-ft.

This mill is supplied by the Hydraulic Race Co. (north tunnel) and has no connection with the 32-foot head noted below.

The Hydraulic Race Co. plans this spring to deepen the north tunnel to the bottom of the upper canal level and to build a hydro-electric plant just below the old locks.

NOTE.— All of the aforementioned plants are supplied by the Hydraulic Race Co. at a 50-foot head and are in no wise dependent on flow in Eighteen-Mile creek, because they merely by-pass water around Locks Nos. 34 and 35 and into the lower level of the Erie canal.

I. and T. Huston.   Cold storage plant.

Have one 22-inch horizontal Leffel wheel, rated at 72 hp., about 10 years old.   Are now using about 50 hp. with head of 18 to 20 feet.

Their right is to the waters of Eighteen-Mile creek, *i. e.*, about 3 square miles drainage area.

Notes on Power Installation

United Paper Board Co.   U. M. Waite, Mill Manager.

Has 32-foot head at canal and 14-foot head below Clinton street.   Former is not in use, but at latter there is an old wheel of 40 to 50 per cent efficiency.   The grant from the Government of 500 sec.-ft. is revocable, as is also the lease with the State for carrying the same.   The mill owners are seeking a 50-year lease, and as soon as a constant flow of 500 sec.-ft. is assured, this company plans to install at the canal for 1,400 hp., utilizing 32-foot head, at an estimated cost of $150,000.   Mr. Waite thought they might install for an additional 500 sec.-ft. through the closed season of navigation, if such were granted.

Lockport Paper Co.   Henry Nichols, Manager.

I did not see Mr. Nichols, but Mr. Gyatt states that this company is already equipped to use 500 sec.-ft. at a 12-foot head.   The installation was new in 1918 and develops an efficiency of 85 to 88 per cent.   About 4 feet additional head could be obtained if the company raised its spillway and acquired some adjacent flat land.

Has an auxiliary steam plant and also buys some power from the city.

Niagara Paper Co.   Mr. Green at head.

Has two horizontal water wheels, 20 years old, with efficiency of perhaps 60 per cent.   Also uses a steam plant and motors, aggregating about 555 hp.

Intends to install new plant when constant flow of 500 sec.-ft. is assured.   Would probably install for an additional 500 sec.-ft. through the winter, if such flow were granted.

Westerman & Co. Calvin Sutliffe, Owner.

Have one 35-inch water turbine in use, about 1½ years old. Also have steam plant. Have no plans at present, but would install for 500 sec.-ft., when assured of same.

Would not install for additional 500 sec.-ft. in winter. Instead of hydro-electric plant, might connect water wheels direct with trains of rollers. Might consider combining head with Indurated Fibre Co. and Electric Smelting & Aluminum Co. for a common power plant.

Indurated Fibre Co. Mr. Bates, Manager, also President of Lockport Mill Owners Association.

31.4-foot head. Have three water wheels in use, a 17-inch Victor, a 32-inch McCormack and a 48-inch McCormack. One is direct connected, grinding pulp, one pumping water and one runs four beaters. 60 to 75 per cent efficiency.

Will put in new installation when constant flow of 500 sec.-ft. is obtained.

Would not install for additional 500 sec.-ft. in winter months. Would probably install its own individual plant.

Electric Smelting and Aluminum Co. Mr. Davis, Manager.

Now uses 30 to 40 hp. (one water wheel, about 50 per cent efficiency) and some electric power.

Has plans for installing two new units to generate 1,800 hp. at an estimated cost of $82,000, when assured of a constant flow of 500 sec.-ft. New installation guaranteed 88 per cent efficiency at 90 per cent load.

Would not install for additional 500 sec.-ft. in winter months, but desires same amount of water the year around.

Newfane Lumber & Mfg. Co. C. G. Evans, Operator

Has one 62-inch vertical turbine about 5 years old, Leffel make, 80 per cent efficiency. Capacity is 250 cubic feet per second. Output now averages 200 hp., 24 hours per day.

Fred Collins & Son. Mill at Newfane.

Have two turbines, both new. One is a 35-inch rated at 110 hp., and the other is 40-inch, rated at 47 hp. Takes 25 hp. to run feed side of mill and about 25 hp. to run wheat side. Use both.

Lockport Felt Co.   Mr. Andrews, Engineer.   Mill at Newfane.

Has two old vertical wheels, size not known.   By rope drive run a 100-kilowatt generator, developing 110 kilowatts.

Western New York Water Co.

Have a power site at Burt, undeveloped.

STATE RIGHTS IN EIGHTEEN-MILE CREEK

The State has obtained certain rights to discharge waste waters from the canal into Eighteen-Mile creek. These rights were appropriated in 1834 and were for the amount wasted into the creek at that time, which, from the smallness of the canal and of the then water-supply, must, therefore, have been small. If it is decided to increase the flow to an amount materially above 375 second-feet without enlarging the channel, appropriations will be necessary for securing the right to flood the lands.

In the recent past there have been times when it was necessary to deliver more than 375 second-feet into Eighteen-Mile creek and in so doing property has been damaged and the owners have brought claims against the State therefor.

An estimate of the cost of carrying out the various improvements on Eighteen-Mile creek is given in the following table, No. 5.

Maps showing the conditions on Eighteen-Mile creek have been prepared, as follows:

Sheets 1, 2, 3 and 4, each entitled, "Map of Eighteen-Mile Creek, County of Niagara, N. Y., made by the State Engineer and Surveyor pursuant to the provisions of chapter 425, Laws of 1918. Examined and approved Feb. 19, 1919, R. G. Finch, Deputy State Engineer and Surveyor."

Sheet No. 5, entitled "Plan and Profile of Eighteen-Mile Creek," etc.

These maps show the location of the channel of the creek, various dams and mills, center line and general cross-sections of the proposed improvements, etc.

NOTE.— The maps accompanying this report may be found in a pocket at the back cover of this volume.

CONCLUSIONS

The investigations made by this Department warrant the following conclusions:

(1) That the present channel of Eighteen-Mile creek will carry a flow of 375 cubic feet per second without overflowing its banks.

(2) That to enlarge the channel of that creek to carry a flow of 500 second-feet without raising the water-surface above that for a flow of 375 second-feet is estimated to cost $31,855.00, and if the capacity of the stream is increased by flooding adjacent lands and making no excavation in the channel, the estimated cost will be $17,250.

(3) That to enlarge the channel of the creek by excavation, to carry a flow of 1,000 second-feet without raising the water-surface above that for a flow of 375 second-feet, is estimated to cost $211,945.00, and if the capacity of the stream is increased by flooding adjacent lands, and making no excavation in the channel, the estimated cost will be $52,957.50.

(4) That with a flow of 375 second-feet in the present channel the total water-power now in use is 2,070 horse-power and the undeveloped power 7,090 horse-power; i. e., 77 per cent of the possible power on the stream is undeveloped.

(5) That if the channel of the stream is improved to carry a flow of 500 second-feet, the undeveloped power will be increased to 10,150 horse-power, or to five times the amount of power now used.

TABLE NO. 1

Computations of Work to Be Done for Discharging 500 Second-feet, if Water-surface Is to Be Retained at Same Elevation as With Present Discharge of 375 Second-feet, Causing No Damage

RANGE No. (1)	Distance (2)	Excavation (3)	Excavation per linear foot (4)	Area needed for spoiling excavation (5)	Work to be done on present channel (6)	Flooded area, if present channel is to be retained (7)
	Feet	*Cu. yds.*	*Cu. yds.*	*Acres*		*Acres*
1.......	11,866	16,261	1.4	6.8	Deepen 1.0 foot.	55.9
14......	3,266	1,089	0.3	1.5	Deepen 0.2 foot	11.3
19......	3,675	7,486	2.0	3.7	Widen 14 feet ..	9.8
23......	1,737	Clean out	Clean out	0.8	Clean out......	4.9
25......	10,096	Clean out	Clean out	4.6	Clean out......	13.7
35......						
Totals.........		24,836	....	17.4	....	95.6

DRAINAGE AREAS:
    Olcott to junction with east branch........... 17 square miles.
    East branch................................. 40 square miles.
    Above junction with east branch..... .... 25 square miles.
    Small branch above canal... .... 3 square miles.

      Total...................... .... 85 square miles.

TABLE NO. 2

Computations of Work to Be Done for Discharging 1,000 Second-feet, if Water-surface Is to Be Retained at Same Elevation as With Present Discharge of 375 Second-feet, Causing No Damage

RANGE No. (1)	Distance (2)	Excavation (3)	Excavation per linear foot (4)	Area needed for spoiling excavation (5)	Work to be done on present channel (6)	Flooded area, if present channel is to be retained (7)
	Feet	*Cu. yds.*	*Cu. yds.*	*Acres*		*Acres*
1.......	11,866	65,043	5.5	13.3	Deepen 4.0 feet.	80.6
14......	3,266	12,338	3.8	3.1	Deepen 2.0 feet.	17.2
19......	3,675	46,686	12.7	13.7	Widen 88 feet...	26.4
23......	1,737	6,369	3.7	8.0	Widen 31 feet...	6.8
25......	10,096	86,751	8.6	27.3	Widen 57 feet...	64.4
35......						
Totals	....	217,187	....	65.4	....	195.4
35......						
55......	Present channel sufficient, if cleaned out............................					100
R. W. & O. R. R. bridge..	Present channel sufficient..					12
Olcott....	Present channel sufficient..					....
Total.	....					307.4

TABLE No. 3

Waste of Water-power at the Present Capacity of Stream (375 Second-feet)

PLANT	Present working head	Undeveloped head	Total head claimed	Horse-power developed at present	Available horse-power at 80 per cent efficiency	Available horse-power undeveloped	Present installation
L. and T. Huston, United Paper Board Co..........	19	13	32	50	1,090	1,040	22-in. wheel, 10 yrs old.
United Paper Board Co.....	11.2	2.8	14	......	480	480	1 old wheel, 40 to 50% efficiency.
Lockport Paper Co..........	11.2	0.9	12.1	380	410	30	For 550 hp.
Niagara Paper Co..........	8.8	0.9	9.7	230	330	110	2 54-in. turbines, 20 yrs. old, 60% efficiency.
Westerman & Co...........	19.4	1.1	20.5	180	700	520	1 35-in. turbine, 80% efficiency.
Indurated Fibre Co........	26	5.4	31.4	800	1,070	270	1 17-in. wheel; 1 33-in. wheel; 1 45-in. wheel; 60 to 75% efficiency.
Electric Smelting and Aluminum Co................	30.1	4.1	34.2	40	1,170	1,130	30-in. Victor wheel and 35-in. Victor wheel not in use; 2 small wheels in use.
Niagara Farmers Co........	0	5	5	0	170	170	None.
Horton Mills..............	0	9.5	9.5	0	320	320	Ruins.
Newfane Lumber and Manufacturing Co., and Fred Collins & Son............	13.4	0.3	13.7	250	470	220	62-in. turbine; 5 yrs. old; 35-in. turbine, new; 40-in. turbine, new.
Lockport Felt Co..........	7.3	1.6	8.9	150	300	150	2 old vertical wheels.
Western N. Y. Water Co....	0	47	47	0	1,600	1,600	None.
Totals................	146.4	91.6	238.0	2,070	8,110	6,040	
Unclaimed power sites......	......	30.9	30.9	......	1,050	1,050	
Totals................	146.4	122.5	268.9	2,070	9,160	7,090	

SUMMARY — Total fall between lower level of canal and Lake Ontario......... 268.9 feet.
　　　　Total horse-power under 268.9 head (80% efficiency).............. 9,160 hp.
　　　　Horse-power developed.................................... 2,070 hp.
　　　　Horse-power undeveloped.................................... 7,090 hp.
　　　　Percentage wasted.. 77 per cent.

TABLE No. 4
Waste of Power at a Flow of 500 Second-feet

PLANT	Present working head	Unde-veloped head	Total head claimed	Horse-power devel-oped at present	Available horse-power at 80 per cent efficiency	Available horse-power unde-veloped	Present installation
L. and T. Huston, United Paper Board Co..........	19	13	32	50	1,450	1,400	23-in. wheel, 10 yrs. old.
United Paper Board Co.....	11.2	2.8	14	0	640	640	1 old wheel, 40 to 50% efficiency.
Lockport Paper Co..........	11.2	0.9	12.1	390	550	170	For 550 hp.
Niagara Paper Co..........	8.8	0.9	9.7	230	440	230	2 24-in. turbines, 20 yrs. old, 60% efficiency.
Westermann & Co..........	19.4	1.1	20.5	180	930	750	1 35-in. turbine, 80% efficiency.
Indurated Fibre Co.........	26	5.4	31.4	800	1,430	630	1 17-in. wheel, 1 32-in. wheel, 1 48-in. wheel; 60 to75% efficiency.
Electric Smelting and Aluminum Co................	30.1	4.1	34.2	40	1,560	1,520	30-in. Victor wheel and 35-in. Victor wheel not in use; 2 small wheels in use.
Niagara Farmers Co........	0	5	5	0	230	230	None.
Horton Mills..............	0	9.5	9.5	0	430	430	Ruins.
Newfane Lumber and Manu-facturing Co., and Fred Collins & Son...........	13.4	0.3	13.7	250	620	370	63-in. turbine, 5 yrs. old; 35-in. turbine, new; 40-in. turbine, new.
Lockport Felt Co..........	7.3	1.6	8.9	150	400	250	2 old vertical wheels.
Western N. Y. Water Co....	0	47	47	0	2,140	2,140	None.
Totals..............	146.4	91.6	238.0	2,070	10,820	8,750	
Unclaimed power sites......	......	30.9	30.9	......	1,400	1,400	
Totals..............	146.4	122.5	268.9	2,070	12,220	10,150	

SUMMARY — Total horse-power under 268.9-ft. head (at 80% efficiency)........ 12,220 hp.
         Horse-power developed at present............................ 2,070 hp.
         Horse-power undeveloped.................................. 10,150 hp.
         Percentage wasted.. 83 per cent.

TABLE No. 5
Estimates of Cost of Enlarging Eighteen-Mile Creek to Carry Flows of 500 to 1,000 Cubic Feet per Second

1.— Estimate of cost to put creek in condition to carry 500 second-feet.
     (A) *By excavating channel.*
25,000 cu. yds. of excavation @ $1.00...................... $25,000
18 acres of land for spoil-banks @ $150...................... 2,700
Engineering and contingencies, 15 per cent..................... 4,155

     Total $31,855

(B) *Cost if land is flooded. No excavation.*

100 acres of land flooded @ $150.............................. $15,000

Engineering and expenses, 15 per cent......................... 2,250

Total $17,250

2.— Estimate of cost to put creek in condition to carry 1,000 second-feet.

(A) *By excavating channel.*

218,000 cu. yds. of excavation @ $.80......................... $174,400

66 acres of land for spoil banks @ $150........................ 9,900

Engineering and contingencies, 15 per cent..................... 27,645

Total $211,945

(B) *Cost if land is flooded. No excavation.*

307 acres of land flooded @ $150.............................$46,050 00

Engineering and expenses, 15 per cent......................... 6,907 50

Total $52,957 50

BOUNDARY LINE REPORTS

SURVEYING AND MONUMENTING PART OF THE DELAWARE‑
SCHOHARIE COUNTY LINE

SURVEYING AND MONUMENTING THE WARRENSBURG–LUZERNE
TOWN LINE AND THE WARREN–SARATOGA COUNTY LINE

SURVEYING AND MONUMENTING PART OF THE ULSTER–GREENE
COUNTY LINE

SURVEYING AND MONUMENTING PART OF THE CAYUGA–WAYNE
COUNTY LINE

EXAMINATION OF THE NEW YORK–NEW JERSEY STATE LINE
MONUMENTS

DELAWARE–SCHOHARIE COUNTY BOUNDARY LINE

State Engineer's Report on the Survey of a Portion of the
Boundary Line Between the Counties of
Delaware and Schoharie

(Chapter 559, Laws of 1918, Section 1)

The following is the report of the field work done by and under
the direction of the State Engineer and Surveyor in locating, establishing and permanently marking upon the ground a portion of the
boundary line between the counties of Delaware and Schoharie.

Chapter 559, Laws of 1918, section 1, reads in part as follows:
" The State Engineer and Surveyor is hereby authorized and
directed to locate, establish and permanently mark upon the
ground the county boundary line between the towns of Gilboa and
Jefferson, Schoharie county, and the towns of Roxbury, Stamford
and Harpersfield, Delaware county, from a point where the counties of Greene, Delaware and Schoharie meet at Utsayantha lake."

In 1914 a survey of the county line between Greene and Schoharie counties was made in accordance with chapter 760, Laws of
1913. In the report of this survey, in Vol. I of the State Engineer
and Surveyor's report for the fiscal year ended September 30, 1914,
there is a description of the line which is now the defined line
between Delaware and Schoharie counties between the points mentioned in the act under which the present survey was made. This
description states " that by chapter 63, Laws of 1788, of the Sen-
ate and Assembly of the State of New York, the counties of Albany
and Ulster were formed and a portion of the boundary line between
them was established as follows: '. . . and running thence to
the head of Kaaters creek or kill, where the same issues out of the
southerly side or end of a certain lake or pond lying in the Blue
mountains; then from thence to a small lake called Utsayantho,
and thence north . . ' By chapter 42, Laws of 1795, Schoharie
county was erected from the counties of Albany and Otsego and a
portion of the dividing line between Albany and Schoharie coun-

NOTE.— The map accompanying this report may be found in a pocket at the
back cover of this volume.

ties was defined as follows: '. . . . beginning at the northwest corner of the manor of Rensselaer and running thence southerly along the westerly line thereof, to the southwest corner of said manor thence westerly on a direct line to the place where the Ulster county line crosses the Schoharie creek thence along the said line to Lake Utsayantho . . .' By chapter 59, Laws of 1800, Greene county was erected from the counties of Albany and Ulster. By chapter 123, Laws of 1801, 'the extent and limits' of the thirty counties of New York state were defined. The boundary line of Greene county is defined in part as follows: '. . . to the southwest corner of the manor of Rensselaerwyck, and a line drawn from thence to the place where the line formerly run from the head of Kaaters creek issuing out of the southerly side or end of a certain lake or pond, lying in the Blue mountains, to a small lake called Utsayantho intersects the Schoharie creek, and westerly by the said county of Delaware.' By the same act the boundary of Schoharie county is defined in part as follows: '. . . along the north bounds of Harpersfield to the said Lake Utsayantho and southerly by a line formerly run from the head of Kaaters creek, where the same issues out of the southerly side or end of a certain lake or pond, lying in the Blue mountains to the said Lake Utsayantho, and by part of the north bounds of the county of Greene.'"

In the act of 1801, in addition to that portion quoted above, it is stated that the boundary of Delaware county is defined in parts as follows: ". . . . and along the southerly bounds of the county of Schoharie to the Lake Utsayantho."

By the survey of 1914 the point on the line between " the end of a certain lake or pond in the Blue mountains " and " the Lake Utsayantho " and in the center of the Schoharie creek was established. By chapter 559, Laws of 1918, under which this present survey was to be made, this point, which is the point where Delaware, Greene and Schoharie counties meet, was designated as the eastern end of the survey. A straight line from this point to Lake Utsayantho is a portion of the line defined in the act of 1801 and is a line forming a boundary between Delaware and Schoharie counties. As to the exact location of the western end of the line at Lake Utsayantho, nothing was found in laws or records to indicate just what point in the lake should be taken. An old map was found in the State Library which showed that in 1787 a line was run from Kaaters creek in the direction of Lake Utsayantho, but this line struck north of the lake. A line was then evidently run from Lake Utsayantho in the direction of Kaaters creek, but this line struck south of the head of the creek. A line was shown on the map as connecting the southern side of the lake with the head of Kaaters creek, but was evidently not run out. After an extensive study of the laws affecting this line and of the lake itself, it was decided that it was evidently meant that the outlet of the lake, as in the case of "the lake or pond in the Blue mountains," was the point to be taken.

The survey pursuant to chapter 559, Laws of 1918, was made by running a random line approximately along the county line from the spillway of Lake Utsayantho to the point at the center of Schoharie creek (as defined above) and then the bearing of the straight line between these points was computed and found to be S 62° 12′ E. Having established the bearing, the county line was then, by proper surveying methods, established on the ground and monuments were placed as follows:

Monument I is on the south side of Lake Utsayantho at the approximate angle in the county line and is 322.7 feet N 62° 12′ W from center of lake spillway. It is 22.5 feet from a blazed pine tree and 9.6 feet from a blazed maple tree.

Monument II is on a hill southeast of the lake, 3,722.7 feet from Monument I. It is 394 feet northeast of a blazed double oak tree and 210.8 feet southeast of a 14-inch blazed oak tree.

Monument III is on the same hill, 1,000.0 feet from Monument II and about ¼ mile northwest of the "Murphy" house. It is 3 feet west of a stone wall, 80.4 feet northwest of a blazed maple tree and 104.5 feet northeast of another blazed maple tree.

Monument IV is 5,756.4 feet from Monument III and is north of the guard-rail on the State highway from Stamford to Grand Gorge, about 1½ miles from Stamford. It is 10.7 feet southwest of a blazed elm tree and 19.4 feet southeast of a blazed cherry tree.

Monument V is 2,523.6 feet from Monument IV and west of a woods. It is 5 feet west of a stone wall, 5.7 feet northwest of a blazed cherry tree and 35.9 feet north of a blazed beech tree.

Monument VI is 5,620.1 feet from Monument V and is located on a hill. It is 142.6 feet from a blazed maple tree and 90.5 feet northwest of blazed double maple trees.

Monument VII is on a hill 2,970.0 feet from Monument VI. It is about ¼ mile northwest of Patrick Moore's house, and 90.9 feet northeast and 115.8 feet northwest, respectively, of blazed maple trees.

Monument VIII is 1,509.5 feet from Monument VII and north of the highway from Stamford to Grand Gorge. It is S 62° 12' E of Patrick Moore's house and 75.8 feet southwest of a blazed maple tree and 62.2 feet northwest of the northeast corner of the parapet wall of a small highway bridge.

Monument IX is on a hill and is 5,820.5 feet from Monument VIII. It is 452.7 feet southeast of a blazed hickory tree and 284.7 feet from a blazed pine tree.

Monument X is 6,366.2 feet from Monument IX and east of a roadway. It is 90.7 feet east of the corner of Cattone Brothers house and 92.0 feet northwest of the corner of another house.

Monument XI is 4,363.8 feet from Monument X and is located on a hill. It is 7.0 feet southwest of a blazed birch tree and 122.0 feet northwest of a blazed triple chestnue tree.

Monument XII is on the west side of the Grand Gorge–Gilboa highway and is 4,017.0 feet from Monument XI. It is 818.6 feet southwest of a blazed chestnut tree and 215.3 feet northwest of a blazed chestnut tree.

Monument XIII is on a high rock ledge of Pine mountain and is 2,039.2 feet from Monument XII. It is 106.7 feet southwest and 122.6 feet northwest, respectively, of blazed oak trees.

Monument XIV is in the woods on Pine mountain, 4,433.8 feet from Monument XIII and 3,628.8 feet from the center of Schoharie creek, the point referred to early in this report. It is about 5 feet west of a wire fence, 17.0 feet southeast of a blazed birch and 44.8 feet north of a hole drilled in a large rock.

All the monuments were made of concrete. Monuments I, III, IV, VIII, X and XII were about six inches square and thirty inches long with a bronze plug in the top of each. These monuments were set in a concrete base that extended about three feet below the surface of the ground. The remaining monuments were cast in place, a piece of stovepipe being used for the form above the ground, and a piece of heavy copper wire being used for the point in the top. All monuments were reinforced.

A report of the field work done in locating, establishing and permanently marking this line upon the ground, a map showing the location, establishment and permanent marking, together with all field-notes, maps and data obtained, have been filed in the office of the State Engineer and Surveyor, and true copies of the completed report and map have been filed in the office of the State Comptroller and in each of the County Clerks' offices of the counties of Delaware and Schoharie.

(Signed)   FRANK M. WILLIAMS,

State Engineer and Surveyor.

WARRENSBURG–LUZERNE TOWN AND WARREN–SARATOGA COUNTY BOUNDARY LINES

State Engineer's Report on the Survey of the Warrensburg-Luzerne Town Boundary Line, Warren County, and the Warren-Saratoga County Boundary Line

(Chapter 561, Laws of 1918)

The following is the report of the work done by and under the direction of the State Engineer and Surveyor in establishing the boundary line between the town of Warrensburg and the town of Luzerne, and the line between Warren county and Saratoga county, from the Hudson river westerly to the easterly line of Hamilton county.

Chapter 561, Laws of 1918, reads in part as follows: "The State Engineer and Surveyor is hereby authorized and directed to locate, establish and permanently mark upon the ground the line between the town of Warrensburg and the town of Luzerne, in Warren county, and the line between Warren county and Saratoga county, from the Hudson river westerly to the easterly line of Hamilton county."

After an extensive search had been made and a complete report received from Deputy Attorney-General Burton H. Loucks, representing the Conservation Commission, the history of the line to be surveyed was available for the use of the State Engineer's Department and the field work was begun during the month of June, 1918. The search and report referred to showed as follows:

" The towns of Luzerne and Warrensburg lie along the east side of the Hudson river and the division line between them is apparently an easterly continuation of the Warren–Saratoga county line referred to above. The legislative acts affecting this line are as follows:

" March 12, 1772, the Colony passed chapter 1534, entitled 'An Act to divide the county of Albany into three counties . . . Be it enacted by his Excellency the Governor and Counsel and the

---

NOTE.— The map accompanying this report may be found in a pocket at the back cover of this volume.

general assembly, and it is hereby enacted by the authority of the same, that the county of Albany shall be henceforth restricted to the bounds and limits following, to wit: On the south, and on the west side of Hudson river by the county of Ulster . . . to the · Mohawk river, thence north until it intersects a west line drawn from Fort George near Lake George, thence east until it intersects a north line drawn from that high falls on Hudson river next above Fort Edward . . . and be it enacted by the same authority, that all the lands lying within this colony to the westward of the county of Albany as by this Act restricted, and to the westward of the north line drawn from the Mohawk river, above mentioned, continued to the north bounds of this province, shall be one separate and distinct county and be called and known by the name of the county of Tryon. And be it enacted by the same authority, that all the lands lying within this colony to the northward of the county of Albany as restricted by this Act, and to the eastward of the county of Tryon and to the westward and northward of the counties of Cumberland and Gloucester shall be one separate and distinct county, and be called and known by the name of the county of Charlotte.'

" March 8, 1773, there was passed chapter 627, entitled, 'An Act to run out the division between the counties of Albany, Tryon and Charlotte . . .' "

On file in the State Engineer's office are manuscript notes, map and survey of Jessup's Upper Patent in 1772. The northerly and southerly bounds are shown as running due east from the Hudson river. The northerly boundary of the tract, some four miles long, is in the immediate vicinity of the Warrensburg and Luzerne town line, and was actually run out with compass the same year the west line from Fort George was established by legislative enactment. There are also on file in the State Engineer's office notes of a survey of the north line of Saratoga county from Hudson river west, by Seth C. Baldwin, dated September 12, 1798. The bearing of this line is here given as S 88° W. Distances from the Hudson river to various streams, mountains and valleys are given, but these apparently have no relation to map No. 386, noted above.

On the first of April, 1775, the assembly of the province passed chapter 1719, "An Act to alter part of the line that divides the counties of Albany, Charlotte and Cumberland." This act trans-

ferred the land east of Hudson river and south of the Fort George
west line from the county of Albany to that of Charlotte. On the
2nd of April, 1775, the name of Charlotte county was changed to
Washington and that of Tryon to Montgomery.

On the 7th of February, 1791, the legislature divided Albany
county into two parts, the northerly portion taking the name of
Saratoga county.

On the 10th of April, 1792, an act was passed, chapter 59, erecting the town of Fairfield with the Hudson river as west bounds,
and north bounds, the west line from Fort George. On the same
date the town of Thurman was described in part as " beginning at
a creek called McAuley's near the south end of Lake George and
running thence on a direct line to the northeast corner of Fairfield;
thence west along the north line thereof to the River Hudson,
thence along the north bounds of Saratoga county to the east line
of Herkimer county, thence along the east line of Herkimer county
to Clinton county."

On April 6, 1808, by act of Legislature the name of Fairfield
was changed to Luzerne. On March 26, 1813, by act of Legislature Washington county was divided, that portion containing
Luzerne and Thurman taking the name of Warren.

On the 12th of February, 1813, by legislative act that portion of
the town of Thurman lying east of the Hudson river and north of
Luzerne was to be called Warrensburg, the remaining portion
taking the name of Athol.

By the revised statutes of 1830, chapter 2, Title 1, this State
was divided into 56 counties, the boundaries of which are accurately given, the boundaries for Warren and Saratoga being the
same as in previous laws. Section 3 of this act reads as follows:
"All lines, which in the foregoing bounds, are described as courses
indicated by the magnetic needle, are respectively to be taken as
the magnetic needle pointed at the several times when such lines
were originally established."

In this same act the bounds of the several towns are given, that
for Luzerne being ". . . on the north by an easterly continuation
of the north bounds of Saratoga county . . ."

From an examination of the foregoing data it is seen that the
problem in hand was to locate on the ground the ancient west

line from Fort George either by finding and identifying marks or
monuments set by some previous and authorized survey or, failing
this, to run in a line from Fort George in accordance with the
magnetic declination of 1772.

Available reliable records of magnetic declination in the
vicinity of the line extended back only as far as the year 1882.
Hence in order to ascertain the approximate probable declination
of the needle at Fort George at the time this line was established
by statute, recourse was had to a long series of observations made
at Albany. The 1905 report of the United States Coast and
Geodetic Survey gives the declination at Albany in the years 1772
and 1905 as 6 degrees 11 minutes west and 11 degrees and 6
minutes west, respectively, or a change of 4 degrees 55 minutes in
the 133-year interval. The same report gives the declination at
Lake George near Fort George in 1905 as 11 degrees 55 minutes
west. On the assumption that the total change during the 133-year
period was the same in amount and direction at Lake George as at
Albany, a value of 7 degrees 0 minutes west is found as the
approximate declination at Fort George in the year 1772.

A portion of the original Fort George, the starting point of
the ancient west line, is still in existence, having been taken over
as a reservation by the State. Construction of the fort was begun
early in 1759, by forces under Gen. Amherst as a base for operations against Fort Ticonderoga and Crown Point. As laid out
it was to have been some 500 feet square with a bastion at each
corner, the ramparts 12 to 18 feet thick, of carefully laid masonry
backfilled with earth and the whole protected by an outer series
of breastworks and rifle pits. The campaign to the north having
been successful, there was no further need for the fort and work
was suspended before December of the same year, a portion of
the rifle pits and the southwest bastion having been completed. (See
Life of Gen. Amherst; also *Set of Plans and Forts in America*,
by Roque; and *Memoirs upon the Late War between the French
and English*, by M. Pouchot.)

Working westward along a line with a true bearing of S 83°
W from Fort George, inquiry and search was made for marks or
other evidence that might lead to the identification of some point
on the true line. The first point of any importance was a very

old blue beech tree with two sets of marks, one set probably eight or ten years old, the other nearly grown over. This tree stands on the west bank of the Hudson river about one and one-third miles below the Stony Creek railroad station. It has been used by local surveyors and generally accepted by the residents round about as being on the division line between the counties of Warren and Saratoga. About a mile farther west, on Hadley Hill at a three-way road junction, a reputed and accepted division point was found where the road work of the towns of Hadley and Stony Creek met. Half a mile west of this there was a property line fence reputed to be on the county line and so recorded in various deeds. About four miles to the west of the Hudson river the line was reputed to lie in the saddle between two mountains, called Round Top and West. Five miles from the river a property line fence according to tradition and deed was on the county line. At little farther on the charred stump of a huge hemlock tree was found. A local resident claimed that before the fire of a few years previous it was plainly marked and was generally accepted as property and county line. About eight miles to the west of the river and on the road leading into Livingston lake there are two marked trees. West of Livingston lake for some distance there are two county lines used, one possibly half a mile north of the other. From here to the Hamilton county line a brief reconnaisance showed more marked trees, and it was evident that the line through these points was far from straight.

The line runs through the rough and wild country forming the Adirondacks. About 75 per cent of its length is through a dense and heavy hardwood forest and with the exception of a few small stretches it is sparsely settled. Owing to the character of the country, it was decided that the most feasible method of determining the relation of the various marks reputed to be on this line, both to each other and to the line from Fort George, was to make the survey by triangulation. Consequently, on June 18, 1918, field work was begun at Lake George. A base line was staked out and its length and true astronomical bearing determined. As the starting point of the whole survey the center of the one completed (southwest) bastion was taken and a connecting random line was run from the base line. A series of triangulation stations

were established on mountain tops from one to two miles apart
and as near as practicable to the probable location of the final
line. Another series of stations was placed somewhat to the north,
angles between the two observed and computations made, working
always toward the west. By the second week in August a base
line had been established along the Hudson river and the work
checked as to distance and bearing.

From a study of the data secured it was found:

That a line drawn through these various points would be very
crooked;

That an average line drawn through them would strike
considerably north of Fort George;

That the bearing from Fort George to any of these points fell
between the extremes of S 82° 18′ W and S 82° 38′ W;

That the bearing from Fort George to the blue beech, previously
mentioned, on the west bank of the river and undoubtedly the
most reliable of all the markers was S 82° 38′ 20″ W;

That the bearing of the north line of Great Lot No. 2, Jessup
Patent, was S 82° 41′ W.

The north boundary of Jessup's Patent was run out in the field
in 1772, the date of passage of the act establishing the "Fort
George west line." The north line of Great Lot No. 2 is parallel
to the north boundary of the Patent and runs from the river easterly
some five or more miles nearly parallel to and about 600 feet north
of the Fort George–blue beech line. The difference in bearing of
these lines, as will be noted, is some three minutes, much less than
might be reasonably be expected.

Thus it is seen from the data that the blue beech tree meets
the requirements of a generally accepted monument and that it
is further checked as having the correct bearing from Fort George
as determined by the best available data. The work of monumenting this line, as thus established, was begun at the Hudson river
and carried easterly about five miles. Triangulation stations
on both sides of the Hudson were permanently marked and tied in
for future use. On September 23, 1918, field work was suspended
for lack of funds.

Monuments were set as follows and because of the densely
wooded country they were well tied in:

Monument No. 1. On the east bank of the Hudson river and west of the highway; a cast-in-place concrete marker 5.2 feet long set 4.2 feet in the ground, 6 inches square at top and 8 inches square at ground-surface, marked on the south, " L ", on the north, " W ", centered with ¾-inch galvanized-iron pipe projecting 3 inches. The northwest corner of Eddy's barn is due south 147 feet; a marked twin pine, N 20° W, 20 feet; a marked twin pine, S 45° W, 38 feet; and a marked four-inch pine, N 85° W, 12 feet (on town line).

Monument No. 2. A 3/16-inch brass bolt, screw head, set in highest point of ledge at summit of ridge and 12,970 feet east of Monument No. 1. A marked 10-inch maple is S 67° E, 19.7 feet; a marked 6-inch oak, N 47° W, 4.2 feet; and a marked 15-inch pine, S 14° E, 26.3 feet.

A cutting was made on the town line entirely across the top of the ridge.

Monument No. 3. A 3/16-inch brass bolt, screw head, set in ledge rock, about 4 feet above and 12 feet to west of center of road leading along west side of Stewart brook and 16,882 feet east of Monument No. 1. The southwest corner of a cellar hole is S 85° 15′ E, 106 feet; the northwest corner of a barn, S 0° 45′ W, 176 feet; a marked 8-inch elm, S 11° 30′ E, 124 feet; a marked 8-inch maple, S 6° 15′ E, 101 feet; a marked 6-inch butternut, S 27° 0′ W, 15 feet; and a marked 3-inch elm, N 82° 30′ W, 26 feet.

Monument No. 4. 18,869 feet east of Monument No. 1; a cast-in-place concrete marker 3 feet long set on rock (2 feet under ground-surface), 4 inches square at top, 6 inches square at surface of ground and centered with a ¾-inch galvanized-iron pipe projecting 3 inches. Not tied in, but signal tripod 10 feet high left in place over monument.

Monument No. 5. 25,997 feet east of Monument No. 1; a 3/16-inch brass bolt set in top of large rock. A 10-inch beech is S 83° 30′ W, 13.5 feet; a 6-inch beech, N 18° 0′ W, 20.9 feet; and a 5-inch maple, N 42° 0′ E, 6.0 feet.

Monument No. 6. 26,212 feet east of Monument No. 1; a 24-inch hemlock tree, with " W " and " L " cut about breast high in north and south sides, respectively.

The line was also cut across ridge about 6,000 feet to east of Monument No. 1.

Using "Map of Lots 1 to 5 Jessup 7755A Patent from Survey by I. C. Wood, Chief Land Surveyor" (Conservation Commission) dated May, 1912, as a means of identification, and from actual measurements made in the field, it is found that the north line of Great Lot No. 2 is 609 feet and 613 feet north of Monument No. 1 and Monument No. 3, respectively.

It is recommended that additional funds be provided at an early date in order that the work already done may be conserved.

A report of the field work done in locating, establishing and permanently marking this line upon the ground, as above described, a map showing the location, establishment and permanent markings, together with all field-notes, maps and data obtained, so far as the work has been completed, have been filed in the office of the State Engineer and Surveyor, and true copies of this report and map have been filed in the office of the State Comptroller and in each of the County Clerks' offices of the counties of Warren and Saratoga, State of New York.

(Signed)   FRANK M. WILLIAMS,
State Engineer and Surveyor.

ULSTER-GREENE COUNTY BOUNDARY LINE

State Engineer's Report on the Survey of Part of the Boundary Line Between the Counties of Ulster and Greene

(Chapter 562, Laws of 1918, Section 1)

The following is the report of the work done by and under the direction of the State Engineer and Surveyor in locating, establishing and permanently marking upon the ground a portion of the boundary line between the counties of Ulster and Greene, as performed during the years 1918 and 1919.

This survey was authorized by chapter 562, Laws of 1918, section 1, which reads in part as follows: "The State Engineer and Surveyor is hereby authorized and directed to locate, establish and permanently mark upon the ground a portion of the north boundary line of the county of Ulster and being a portion of the south boundary line of the county of Greene and known as the north boundary line of Great Lot number eight, Hardenburgh Patent."

The survey was started in August, 1918, and continued until December 13, 1918, when work was closed for the winter. By chapter 600, Laws of 1919, an additional appropriation was made and the work carried to completion.

By chapter 59, Laws of 1800, Greene county was erected from the counties of Albany and Ulster. By chapter 123, Laws of 1801, "the extent and limits" of the thirty counties of New York state were defined. Chapter 46, Laws of 1812, states: "That from and after the passing of this Act, the division line between the counties of Ulster and Greene shall begin at the point where the division line between the counties of Ulster and Delaware intersects the line run for the north-easterly bounds of Great Lot number eight, in the Hardenburgh Patent, then southeasterly along the said line until it intersects the line run by Jacob Tremborn, Junior, in the year one-thousand eight hundred and eleven for the division between the counties of Ulster and Greene, thence along the last mentioned line . . . ."

NOTE.— The map accompanying this report may be found in a pocket at the back cover of this volume.

This law and also chapter 562, Laws of 1918, states that the boundary line is a portion of the north boundary line of Great Lot number eight, Hardenburgh Patent.

The partition deed of November 14, 1749, Liber 14, page 538, Office of the Secretary of State, describes lot number eight as follows: "Lot number eight bounded southerly by lot number seven, northerly by a line to be drawn from monument number fourteen on the said branch (East branch of Delaware river) to monument number four on the east of the patent, being a heap of stones around a red Oak tree N. 4 standing forty chains to the southward of Cornelius Van Keuren's house and thirty chains to the southward of the Sawkill."

The records in the County Clerks' offices of Delaware and Ulster were searched for information concerning monuments fourteen and four. Records were found deeding property from these monuments. An old map was found in the State Library which showed monument four and its position in relation to the Sawkill and Cornelius Van Keuren's home. An old map was also found in the possession of Sheldon Vreddenburgh of Zena, N. Y., which also showed monument four and its connection with his property. With the above information a portion of the property near monument four was run out and the position of the monument located on the ground. This position was also checked in relation to the Sawkill and the VanKeuren house. As all these methods located the monument in approximately the same place, there was no doubt but that the correct position of the monument had been found.

The position of monument fourteen was not so difficult to find, as old stone walls that were property lines were easily found and these pointed to the position of the old monument on the river bank. An old dead stump about fifteen feet high and four feet in diameter, which showed many old blazes, was found standing in the west wall. On the river bank a number of large stones were found and some small trees were growing up among the stones.

This survey was made by running a random line from monument fourteen to monument four. The line was first put through from mountain top to mountain top. The length of the line from the three-county point to the Ox Clove was determined by chaining;

for the remainder of the distance, stadia was used. The total distance of line run was about twenty-nine miles. After the length of the line had been determined, a straight line was computed from monument four to monument fourteen, and the true bearing of this line was found to be N 62° 48′ W. Having established the bearing of the county line, the line was thus established on the ground and monuments were placed, as follows:

Monument I is at the approximate corner of the towns of Lexington and Hunter, Greene county, and the town of Shandaken Ulster county. It is on the west side of a road in a private park about 0.6 mile north of Chichester. It is about 57.3 feet from the north edge and 58.9 feet from the south edge of a small summer house with artistic water-wheel, and about 70 feet west of Ox Clove creek.

Monument II is near the edge of a woods, 587 feet from Monument I. It is 64.0 feet north of a 10-inch bass wood tree, 55.5 feet east of a 10-inch basswood tree and 25.3 feet south of a 12-inch twin apple tree.

Monument III is on a mountain, 5,062.4 feet from Monument II. It is 16.0 feet northeast of a 20-inch maple tree, 16.2 feet northwest of a 7-inch beech tree and 3.8 feet northeast of a 12-inch ash tree.

Monument IV is 6,954.0 feet from Monument III, east of Forest Valley highway and about 100 feet south of Bridget Ennist's house. It is 42.9 feet northeast of a 8-inch locust tree, 31.3 feet south of a 7-inch locust tree and 23.6 feet southwest of a 10-inch locust tree.

Monument V is 6,508.9 feet from Monument IV, east of Peck Hollow road and about 200 feet south of Herbert Yotes' house. It is 21.2 feet northeast of a 3-inch maple tree, 25.8 feet southeast of a 4-inch maple tree and 24.9 feet north of a twin apple tree.

Monument VI is on a mountain, 5,953.7 feet from Monument V. It is 11.6 feet north of a 4-inch maple tree, 2.7 feet southeast of a 8-inch maple tree and 8.6 feet southwest of a 4-inch maple tree.

Monument VII is in Bushnellsville, 7,389.2 feet from Monument VI, west of Bushnellsville–West Kill highway and north of Bushnellsville creek. It is 90.7 feet northeast of the northeast corner of Andrew White's barn, 60.8 feet southwest of a 8-inch maple tree and 61.8 feet northwest of the northwest corner of a bridge over Bushnellsville creek.

Monument VIII is 8,329.6 feet from Monument VII, east of highway, and about 2,000 feet north of Wm. Redmond's house. It is 51.8 feet southeast of a 5-inch apple tree, 35.8 feet east of a 10-inch apple tree and 31.3 feet northeast of a 12-inch apple tree.

Monument IX is on a mountain, 3,174.1 feet from Monument VIII and is near the intersection of the counties of Ulster, Greene and Delaware. It is 50.1 feet southeast of a 15-inch maple tree, 60.6 feet west of a 24-inch birch tree and 78.5 feet northeast of a 18-inch oak.

All the monuments were made of concrete. Monuments I, II, IV, V, VII, VIII and IX are about 4½ inches square at the top, 5½ inches at the bottom, and three feet long, with a copper wire point in the top of each. These monuments were set in a concrete base that extended about three feet below the surface of the ground.

The remaining monuments were cast in place, a piece of stove-pipe being used for the form above the ground and a piece of copper wire being used for the point in the top of these monuments. All monuments were reinforced with iron rods.

It should be noted that Monument I does not indicate the point where that line which, extending northeasterly, forms the boundary line between Ulster and Greene counties intersects the line established, for the funds were not sufficient to trace out that line, but said line would intersect the established line at or near Monument I.

A report of the field work done in locating, establishing and permanently marking this line upon the ground, a map showing the location, establishment and permanent markings, together with all field-notes, maps and data obtained, have been filed in the office of the State Engineer and Surveyor, and true copies of the completed report and map have been filed in the office of the State Comptroller and in the County Clerks' offices in the counties of Ulster and Greene.

(Signed)   FRANK M. WILLIAMS,
State Engineer and Surveyor.

CAYUGA–WAYNE COUNTY BOUNDARY LINE

State Engineer's Report on the Survey of a Portion of Boundary Line between the Counties of Cayuga and Wayne

(Chapter 484, Laws of 1919)

The following is the report of the field work done by and under the direction of the State Engineer and Surveyor in locating, establishing and permanently marking upon the ground the line between the town of Conquest in the county of Cayuga and the town of Savannah in the county of Wayne, as performed during the summer of 1919.

The basis of the survey consisted of two approved marks, one a stone monument at the northerly end of the County Line road and on the northerly side of the easterly and westerly road leading to Spring lake, the other a 12-inch cedar post set in a pile of stones and known as the "Hiram Sibley corner." These two marks are approved and have been used by local surveyors for many years.

The line was run through these two marks south to the center of the Seneca river, thence down the center of the Seneca river to the "South channel," the center of which is the southerly line of the town of Conquest and the northerly line of the town of Mentz, both in Cayuga county. The line was produced northerly to the easterly and westerly line between the town of Butler and town of Savannah, both in Wayne county. This line produced northerly passed through large trees which are to the north on this line and in general satisfied all conditions in the field.

Many authoritative records and maps were consulted and after giving due consideration to all available data the line was established and marked as shown on the accompanying map, entitled, "Boundary Line between the Towns of Conquest, Cayuga County, and Savannah, Wayne County. Survey and map made by the Department of the State Engineer and Surveyor, as authorized by Chapter 484, Laws of 1919."

NOTE.— The map accompanying this report may be found in a pocket at the back cover of this volume.

Ties and the permanent monuments were set as shown on the map. All bearings refer to the true north, which was established by observation on Polaris, made on September 24, 1919.

The field-notes showing information in connection with the field work of making the survey are filed in the office of the State Engineer and Surveyor, Albany, New York.

This report is filed in the office of the Comptroller of the State of New York, in the office of the State Engineer and Surveyor and in the offices of the County Clerks and County Treasurers of the counties of Cayuga and Wayne, as required by chapter 484, Laws of 1919.

(Signed)   FRANK M. WILLIAMS,
State Engineer and Surveyor.

NEW YORK–NEW JERSEY BOUNDARY LINE EXAMINATION

Joint Report of the Engineers on the Examination of the
Monuments Marking the Boundary Lines Between
the States of New York and New Jersey,
Made in May, 1919

The undersigned engineers, designated by the State Engineer
and Surveyor of New York and the New Jersey State Board of
Commerce and Navigation to make an examination of the monuments marking the boundary lines between the states of New York
and New Jersey, have the honor to submit this report and detailed
description of the monuments.

CLASSIFICATION OF MONUMENTS

In Raritan Bay

The monuments marking that part of the New York and New
Jersey boundary line lying across lands under water in Raritan
bay consist of the Romer beacon or lighthouse, the Permanent
monument, the Great Beds lighthouse and the Morgan Range
beacon. (See photographs.)

In Arthur Kill and Kill von Kull

The range monuments marking that part of the New York and
New Jersey boundary line lying across lands under water in
Arthur kill and the Kill von Kull are of granite, four feet long,
having the tops dressed eight inches by eight inches in cross-section
and for a distance down of about eight inches. On the north face
generally of each are cut the letters " N. Y." and " N. J." and on
the south, or opposite face, the letters " B. M." (meaning "Boundary Monument") and the figures " 1889." One-quarter-inch
grooves are cut at right angles across the top parallel with the sides,
one groove entirely across the top and other other only partially
across. They project about four to eight inches above ground.

Between Hudson and Delaware Rivers

The monuments marking that part of the New York and New Jersey boundary line extending from the Hudson river to the Delaware river may be divided into five classes and consist, first, of the original mile monuments erected in 1774; second, the new mile monuments; third, the railroad monuments; fourth, the wagon road monuments; and fifth, the terminal and witness monuments, all of which were set by the Joint Commission in 1882, except the original monuments of 1774, which were reset, when necessary, by said Commission at said time.

The monuments of the first class, the original mile monuments, are composed for the greater part of red sandstone posts, dressed with eastern and western upper angles rounded and are generally fifteen inches wide in the direction of the boundary, seven inches thick and project above ground about twenty inches. The remainder are of rough, irregularly shaped rock of different material, all having the name of the State cut on the appropriate side of the stone and also the number of the mile distant from the eastern terminus of the line.

The monuments of the second class, the new mile monuments, are of granite, four feet long, the tops dressed generally six inches by six inches in cross-section and for a distance down of six inches. Upon the north side are cut the letters " N. Y." and on the south side the letters " N. J." and upon the east side the number corresponding to the number of miles distant from the eastern terminus of the line. One-quarter-inch grooves are cut at right angles across the top parallel to the sides; one groove shows the direction of the boundary and the other is perpendicular to it. They project above ground generally about six inches. They are set east of and generally adjacent to the old mile stones.

The monuments of the third class, the railroad monuments, are generally similar to the mile monuments, except that they are not marked with numbers.

The monuments of the fourth class, the wagon road monuments, are of granite and are four and one-half feet long; the tops are dressed six inches by twelve inches in cross-section and for a distance down of twelve inches. They are marked in the same manner as the railroad monuments and project generally about twelve inches above the ground.

The monuments of the fifth class, the terminal and witness monuments, are as follows, viz.:

The monument at the eastern terminus is a large block of trap-rock, seven feet six inches long, three feet two inches high and about four feet thick. It is located at the foot of the Palisades and about six inches above storm tides of the Hudson river. It is marked with a groove upon its perpendicular eastern face for its full height, at a distance of two feet south from its northerly end, and is further marked with the words "Latitude 41° North" and on the north side of the groove the words "New York" and on the south side thereof the words "New Jersey." (See photographs.)

The monument at the western terminus of the line is of cut granite, two feet four inches long, one foot four inches wide, and projects now only one foot five inches above the surface of the rock in which it is embedded. It is marked upon its top surface with one-quarter-inch grooves, showing the direction of the lines of the three states which meet there, and within the surface bounded by the lines the initials of the respective states are cut. The north side of the stone is further marked with the words "Tri States Monument." Each of these terminal monuments has a witness, or reference, monument located in the most suitable place nearest to it. (See photographs.)

LINE MONUMENTS BETWEEN HUDSON AND DELAWARE RIVERS
Examination

The examination of the boundary from the Hudson river to the Delaware river was begun at Station rock on the Hudson river on May 5, 1919. The monuments, with few exceptions, hereinafter noted, were found in good condition. Some of the original mile monuments set in 1774 are gone or lie flat near the granite monuments set in 1882 to further preserve the mile points. Several of the road monuments are covered through raising of grade or other construction and one has three feet of its length exposed and leans against a reservoir wall. Most of the monuments are chipped at the corners and edges, and a few lean slightly, though still firmly imbedded.

The policy of informing residents concerning the location of the monuments and impressing them with the importance of preserving the same, and the practice of changing the names of the title owners in the detailed descriptions of the monuments, to conform with transfers, have been followed consistently. Where the old witness points have disappeared, new ones have been chosen and noted. Where access to a monument is difficult, notes giving directions have been added at the end of the detailed description. These descriptions follow as a part of this report. Considerable time was lost by the lack of topography on the official boundary maps, especially new roads, probably laid out since those maps were made. This information, we believe, may be obtained from official records and should be plotted and placed on the tracings prior to the next inspection.

Photographs were taken of the terminal and witness monuments and copies of these accompany this report.

Recommendations

For the proper preservation of the boundary between the Hudson river and Port Jervis on the Delaware river, we desire to make the following recommendations:

(1) That the latitude inscription on the eastern terminal monument at the foot of the Palisades be recut, as it is nearly obliterated, and that the pole and sign-board be painted.

(2) That a sign be placed at the eastern witness, or reference, monument, warning against chipping or defacing this monument, as several large pieces have been broken from the edges in a shameful manner. (See photograph.) Also that the Inter-state Palisades Park Commission, now opening paths in this section, be requested to open a path along the New York–New Jersey line from this monument to the boulevard so that the same may be readily reached by the citizens of both States.

(3) That monument No. 15, covered with road material, be raised and reset in concrete.

(4) That monument No. 20, situated on the slope of a hillside be reset in concrete.

(5) That monument No. 23, which is on a sloping hill, loose and about ready to fall, be reset in concrete.

(6) That a new monument be set in concrete in place of No. 43, which is under cement curb and sidewalk.

(7) That monument No. 44 be raised to grade of railroad and set in concrete.

(8) That monuments Nos. 52 and 53 be moved to the new roads near-by, as roads on which they are now located are abandoned.

(9) That monument No. 96 be relocated and reset in concrete, as it is now exposed and leaning against a reservoir wall.

(10) That monuments Nos. 113 and 115, now covered, be raised and set in concrete.

(11) That the official maps showing the monuments between the Hudson and the Delaware be brought down to date.

LINE MONUMENTS IN RARITAN BAY AND RANGE MONUMENTS IN ARTHUR KILL, KILL VON KULL, NEW YORK BAY (UPPER) AND HUDSON RIVER

Examination

After completing the examination from the Hudson to the Delaware river your engineers proceeded immediately to Perth Amboy for the inspection of the monuments lying in Raritan bay, Arthur kill and Kill von Kull. Practically all these monuments are in good condition. A few we were unable to find. Owing to the rapid upbuilding of this territory, principally by industrial plants, a considerable number of these monuments are surrounded by or inside buildings. Also quite a few are several feet below grade. In such cases a wooden box, terra-cotta pipe or other device with a wooden cover surrounds the stone and extends to a point near the present surface of the ground. In a few instances we found the covers removed and the holes filled with debris. These covers being of a very temporary character and the grade having been established, we recommend that these monuments be raised to grade and, if possible, where now located among or within buildings, a new point in the range be chosen, from which sights may be readily taken. These are enumerated in our list of specific recommendations which will be found later in this report. As on the north boundary, we found ourselves much inconvenienced by the ancient topography on the official charts.

No inspection was made of the section from New York bay to

Station rock on the Hudson river, as we had no description of these points. Since returning from the field, however, we have made an examination of the history of the boundary and found in the "Report of the Riparian Commissioners of New Jersey for the year 1891" a detailed description of all range points on this portion of the line. They consist of crosses in rocks, church spires, chimneys, public buildings, U. S. Coast and Geodetic Survey monuments, etc. From the best information obtainable it appears these have never been examined since they were established. In view of the fact that this is the most valuable section of the whole territory we recommend that an inspection be authorized and data procured as to amount of work necessary to mark this portion of the boundary to conform with the remainder. An estimate of cost should be included.

Recommendations

For the better preservation of the boundary from Raritan bay to Station rock on the Hudson river we respectfully submit the following specific recommendations:

(1) That an examination be authorized of the section between New York bay and Station rock on the Hudson river.

(2) That the Permanent monument and the Morgan beacon be repainted.

(3) That monuments Nos. 4, 6, 8, 14, 20, 23, 24, 47, 48, 49 and 51 be raised to grade and where situated in a building or surrounded by buildings, a more favorable location be chosen, if practicable.

(4) That monument No. 34 be raised and set in concrete (or a new one set in this range) and located by angles and distances from New York city and U. S. Coast Survey monuments near-by.

(5) That monuments Nos. 41, 42, 43 and 44 be reset in concrete in the roadbed of the B. & O. R. R.

(6) That a new monument be set in concrete near the highway at the edge of the meadow in place of No. 45, which is covered.

(7) That a new map showing topography (except contours) of all the section from Raritan bay to Station rock on the Hudson be compiled in sections of suitable size, marking the names of roads or streets leading to monuments and the names of industrial plants within which monuments are situated. We suggest a scale of one inch to one thousand feet for this map.

To do all the above work there should be a combined appropriation of $10,000, or $5,000 from each State.

Respectfully submitted,

HENRY J. SHERMAN,
Engineer for New Jersey.

H. F. EAGAN,
Engineer for New York.

Dated June 25, 1919.

MONUMENTS ACROSS LANDS UNDER WATER IN RARITAN BAY

A detailed description follows of the location and condition of the monuments, structures and buoys marking that part of the New York and New Jersey boundary line lying across lands under water in Raritan bay.

Romer Lighthouse or Beacon

In good condition.

Permanent Monument

Needs repainting; otherwise in good condition.

Morgan Range Beacon

Consists of a triangular steel tower, about 56 feet high, set on three concrete piers, with a large rectangle of steel at the top as the beacon; all in good condition. It is located on the highland about one mile south of South Amboy, N. J., and about 1,000 feet north of Morgan Station on the New York and Long Branch R. R., on land of the Otis Sand Lime Brick Co., back of their brick manufactory. It is in the prolongation of the line between the Romer lighthouse and the permanent monument. Underneath the rectangular beacon there is an 8-inch by 12-inch granite monument, the southeast corner of which is slightly chipped; otherwise it is in good condition. The steelwork needs repainting.

Great Beds Lighthouse

In good condition.

NOTE.— For the report of the Commission appointed to locate the boundary line in Raritan bay, Arthur kill, Kill von Kull, New York bay, and Hudson river, see the report of the State Engineer and Surveyor for 1896, pages 361–5. For maps of these same sections of the line, see the report of the State Engineer and Surveyor for 1900, facing pages 252 and 254.

MONUMENTS ACROSS LANDS UNDER WATER IN ARTHUR KILL AND KILL VON KULL

A detailed description follows of the location and condition of the range monuments marking that part of the New York and New Jersey boundary line lying across lands under water in Arthur kill and Kill von Kull.

Monument No. 1

This monument is located at Perth Amboy, N. J., in the rear of an old blacksmith shop fronting on Front street and on land of I. B. Eissenberg. It is 3 feet from the doorway, 52.9 feet south of the southeast corner of the brick building of Schantz & Eckert, now used as a storehouse, and 29.0 feet southwest of the southwest corner of the brick portion of the foundry. The said blacksmith shop adjoins the marine railway of Gray & Sons on the north. This monument has its top flush with the surface of the ground. Its northeast and northwest corners are broken and its easterly and westerly edges chipped; otherwise in good condition.

Monument No. 1-A

Quoting from the Engineers' joint report of October 8, 1913:

" This monument, granite, 8 inches by 8 inches by 4 feet long, lettered ' N. Y.–N. J.' on its northerly side and ' No. 1-A ' on its southerly side, is located near the northerly line of Front street in Perth Amboy, N. J. It is 43.7 feet southwesterly at right angles from the southwesterly side, or line, produced of Hartmann's Hotel, 0.5 foot northeasterly at right angles from the northeasterly side, or line, produced of house No. 277 Front street and 2.26 feet easterly from the easterly corner of said last mentioned house No. 277.

" It was set August 11, 1913, on range, or in line, between monument No. 1 and the Great Beds lighthouse and is distant 109.01 feet northerly on said range from said monument No. 1."

This monument is covered with about two inches of earth and has its southerly edge slightly chipped; otherwise in good condition.

PERMANENT MONUMENT, RARITAN BAY

MONUMENT UNDER MORGAN RANGE BEACON

EASTERN TERMINAL MONUMENT, "STATION ROCK"

GENERAL VIEW, STATION ROCK AND SIGN-POST, NEAR THE HUDSON RIVER

EASTERN WITNESS MONUMENT, NORTH SIDE

NEW JERSEY
—
BOUNDARY
MONUMENT
—
1882
—
ABRAHAM BROWNING,
THOMAS N. M'CARTER,
GEORGE H. COOK,
COMMISSIONERS.

489 FEET WEST
FROM STATION ROCK
E. A. BOWSER
SURVEYOR

EASTERN WITNESS MONUMENT, SOUTH SIDE

NEW JERSEY

BOUNDARY
MONUMENT

1882

Abraham Browning
Louis W. Carr
Benere R. Cook
COMMISSIONERS

E. A. Bowser

WESTERN WITNESS MONUMENT

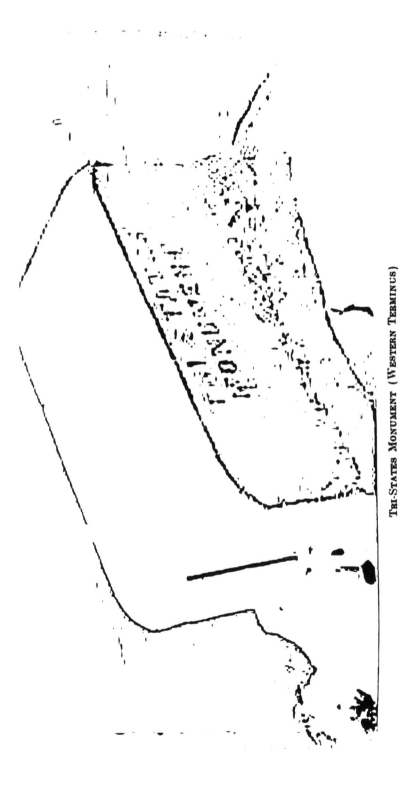

TRI-STATES MONUMENT (WESTERN TERMINUS)

Monument No. 2

This monument is located at Tottenville, S. I., on land of Jere Johnson. It is about 52 yards north of the line fence between lands of Jere Johnson and G. T. Brewster, and about 32 yards easterly from high-water line of the sound on a fairly steep westerly slope. At the monument the following bearings and distances were taken: N 40° E, 25.5 feet to a large red oak tree; N 38° W, 56.5 feet to a large dead twin oak tree, and S 15° W. 44.5 feet to a black oak tree. This monument is covered with about two inches of soil washed from the hillside and has its westerly corner slightly chipped; otherwise in good condition.

Monument No. 3

This monument is located at Tottenville, S. I., on land of Frederick Baxter, 77.7 feet northwest of the northwest corner of Benjamin Williams' house and 88.5 feet, about due west, from the southwest corner of Frederick Baxter's house. This monument has three of its corners and two of its edges slightly chipped and leans a trifle to the west; otherwise in good condition. It is set in concrete.

Monument No. 4

This monument is located at Tottenville, S. I., on land and premises of Alfred Baxter, now occupied by George D. Emmons. It is 27.7 feet northwesterly from the northeasterly corner and 21.23 feet northeasterly from the northwesterly corner of the brick foundation of said Baxter's house; also 137.23 feet easterly from monument No. 3 and about 17 feet southerly from a large beech tree. It is about 3 feet below the present surface of the ground and protected by one length of 15-inch vitrified pipe, which is covered with a two-inch spruce cover, partially decayed, over which there is about eight inches of earth. Its northerly corner is slightly chipped; otherwise in good condition.

Monument No. 5

This monument is located at Tottenville, S. I., on lands of Adeline Dorcey, about 150 feet east of the road leading to the Perth Amboy ferry. At the monument the following bearings and distances were taken: S 20° E, 50.1 feet to the northeast corner of double house occupied by Mrs. Murphy and Mrs. Pryor; S 65° W, 76.2 feet to the southeast corner of S. N. Boderson's barn; N 10° E, 17 feet to wild cherry tree, and N 80° E, about 12.0 feet to a line fence. This monument has its corners and easterly edge slightly chipped; otherwise in good condition.

Monument No. 6

This monument is located at Tottenville, S. I., on land now a shipyard, belonging to Harry Cossey. It is 156 feet northwesterly at right angles from the northwesterly side of Cossey's corrugated iron sawmill; 9.8 feet southwesterly from the northeasterly side, or line, of said Cossey's sawmill produced; about 129 feet southwesterly from his northeasterly bulkhead; 10.5 feet east of end of near rail of siding, and 8.8 feet east of east gage line. It is about 3 feet below the present surface of the ground and protected by one length of 12-inch vitrified pipe, covered with a 5-inch yellow pine cover, over which there is about 12 inches of earth. It is in good condition, but should be reset to grade and set in concrete.

Monument No. 7

This monument is located at Tottenville, S. I., on land of Tottenville Shipyard Co., 82 feet westerly from Fisher avenue and 15 feet northerly from remains of an old fence. At the monument the following bearings and distances were taken: S 80° E, 36.0 feet to a large willow tree; N 50° E, 61 feet to another large willow tree; 43.8 feet south of southwest corner of blacksmith shop; 62.7 feet southeast of northwest corner of said shop, and four paces north of ditch. This monument has its easterly and westerly edges slightly chipped; otherwise in good condition.

Monument No. 8

This monument is located at Tottenville, S. I., on land of the Atlantic Terra Cotta Co. At the monument the following bearings and distances were taken: N 10° E, 63.4 feet to the southwest corner of the office building; N 60° E, 7.0 feet to the northwest corner of the pressing building; S 10° E, 51.6 feet to the northeast corner of the dwelling house of the Company and 8 inches east of west face of pressing building. This monument is about in the center of the roadway leading to the office and about 3 feet below the present surface. It is protected by an 8-inch tile pipe 2 feet in length with cast-iron cap about 8 inches below surface. This monument was otherwise in good condition.

Monument No. 9

This monument is located about one-half mile north of Kreischerville, S. I., opposite Sewaren hotel, N. J., 1 foot south of a line fence, now nearly gone, between lands of Powell and formerly Charles Hughes, now Floyd S. Corbin; about 275 feet from Arthur kill and 25 feet west of drive to Mr. Corbin's house. At the monument the following bearings and distances were taken: S 10° W, 140 feet to a wild cherry tree 6 inches in diameter; N 80° W, 129.0 feet to a pear tree, and N 10° E about 104 yards to the southeast corner of Corbin's old house. This monument was found to be in good condition in every respect.

Monument No. 10

This monument is located at Sewaren, N. J., on land of C. A. Cuppia, 7 feet inside of hedge fence along Holton street and 42 feet inside of hedge fence along Cliff road. At the monument the following bearings and distances were taken: N 20° W, 49.7 feet to the southeast corner of Cuppia's house, and S 80° W, 10.2 feet to a horse-chestnut tree. This monument has its edges slightly chipped; otherwise in good condition.

Monument No. 11

This monument is located at Sewaren, N. J., on land of F. Van Syckle on the north side of the drive leading to his house, about half-way up the slope of said drive and just under the sod. At the monument the following bearings and distances were taken: S 65° E, 20.13 feet to the southwest corner of southerly gate-post just under cap at entrance to drive; N 60° E, 23.92 feet to the northwest corner of northerly gate-post just under cap at entrance to drive, and 38.28 feet northerly to the line fence between lands of F. Van Syckle and Charles Ballard. This monument has its northwest corner slightly chipped; otherwise in good condition.

Monument No. 12

This monument is located on the southeast side of Cliff road in a line of maple trees, about one-half mile north of Sewaren. At the monument the following bearings and distances were taken: S 60° E, 144.4 feet to a large black oak tree, standing alone, near the line between upland and meadow; N 30° W, 105 yards to a semaphore signal, "A 172", on the right of way of the C. R. R. of N. J., and S 35° W, 33.0 feet to a maple tree. This monument has its northeast and northwest corners broken and southeast and southwest corners slightly chipped; otherwise in good condition. It is set in poor concrete.

Monument No. 13

Quoting from the Engineers' joint report of October 8, 1913:

"After quite an extensive survey connecting monuments Nos. 12, 14, 15 and U. S. H. L. monument 'M' on Tufts Point and calculations therefrom, the theoretical, or recorded, position of monument No. 13 was found to be on land of the Philadelphia & Reading R. R. Co. among the tracks usually occupied by loaded coal cars at Port Reading, N. J.

" On September 4, 1913, an attempt was made to find and locate monument No. 13 on the ground by means of the above survey and calculations and by direct measurements over the coal cars from our nearest survey station, with the result that it was not found, in consequence of its theoretical, or recorded, position being in or near the center of one of the many tracks fully occupied by loaded coal cars, which made the question of a further search by digging prohibitive.

" In view of this fact it was deemed advisable to set another stone on the same range with Nos. 13 and 15 and designate it as 13-A."

Monument No. 13-A

Quoting from the Engineers' joint report of October 8, 1913:

" This monument, granite, eight inches by eight inches by 3.3 feet long, lettered 'N. Y.–N. J.' on its northerly side and 'No. 13-A' on its southerly side, is located at Port Reading, N. J., on land of the Philadelphia & Reading R. R. Co. It is in the same line, or range, with monument No. 15 and the theoretical, or recorded position of monument No. 13 and distant 765.95 feet northwesterly therefrom. It is 27.65 feet southeasterly from an iron pump, 36 feet northeasterly from the northeasterly corner of a frame shack and 99.4 feet westerly from the southwesterly corner of another shack.

"It was set September 8, 1913. . . .

"At the monument the following bearings were taken: N 75° E to chimney of the American Agricultural Chemical Co. (Liebig Works) and S 82° E to the northerly one of two water towers on land of the Philadelphia & Reading R. R. Co.

"The theoretical, or recorded, position of monument No. 13 as used above is determined by computations by Mr. William B. Moss, Civil Engineer, of New York city (based on the U. S. C. S. station 'Bogart') and our survey connecting monument Nos. 12, 14 and 15 and U. S. H. L. monument 'M' on Tufts Point."

Monument 13-A is 4 feet southeast of electric light pole No. 40 and near a colony of shacks. This monument has its corners and three of its edges slightly chipped; otherwise in good condition. To reach it take first road to east from main road, north of Port Reading P. O.

Monument No. 14

This monument is located in township of Woodbridge, N. J., about one-half mile west of the borough of Roosevelt on land of the Philadelphia & Reading R. R. Co. At the monument the following bearings and distances were taken: S 5° E, 76.9 feet to the northeast corner of the storehouse of the creosoting plant; N 85° E, 235.5 feet to a steel rail set on end in the ground as a monument at the northeast corner of land of the Port Reading R. R. Co., which corner is 37.0 feet southerly from the centre of a new street, and S 15° E, 466.3 feet to the northwest corner of the brick office building of the creosoting plant. This monument now stands 3 feet below the present surface of the filling and is protected by a yellow pine box made of 4-inch plank. It has three of its corners chipped and easterly corner broken; otherwise in good condition. It should be raised to grade and set in concrete.

Monument No. 15

This monument is located about one-half mile northwest from Rossville, S. I., on land of the Oakland Chemical Co., about 407 feet on range line from the kill and 32.0 feet on range line from the line between upland and meadow. At the monument the following bearings and distances were taken: N 80° E, about 160 feet to the northwest corner of the concrete office building of the company, 72.9 feet north of angle in high fence of company; S 35° W, about 85 paces to the northeast corner of C. C. F. Disoway's house, and N 45° W, 42.0 feet to a large tree. This monument has its northwest, southwest and southeast corners slightly chipped; otherwise in good condition.

Monument No. 15-A

This monument, 6 inches by 6 inches, granite with copper plug in centre, was located about one-half mile northwest from Rossville, S. I., on land of the Oakland Chemical Co., in range with monuments Nos. 13 and 15 and distant 200 feet southeasterly from No. 15 and about on line between Chemical Co.'s water-tower and C. C. F. Disoway's house. This monument was removed in 1916 in consequence of the grading, etc., for a new kiln building.

Monument No. 16

This monument is located at Rossville, S. I., on land of Mrs. Hattie Lyon, 25 feet west of the line fence between lands of Mrs. Lyon and Miss Mary Cole and about 65 feet southwest of high-water line of the kill, in swampy ground and set about 6 inches above the surface. At the monument the following bearings and distances were taken: S 50° E, 27.2 feet to a thorny locust tree; N 20° W, 36.2 feet to an elm tree, and S 80° W, 86.5 feet to a large willow tree. This monument has its southerly edge chipped and top chiseled; otherwise in good condition.

Monument No. 17

This monument is located at Rossville, S. I., on land of Miss Mary Cole, 2.3 feet east of fence which separates her house lot from pasture and 87.0 feet along said fence from the Shore road and about 46 paces west of watercourse. At the monument the following bearings and distances were taken: N 85° W, 8.5 feet to a pear tree, and S 20° E, 45.8 feet to a small apple tree. This monument has three of its corners and three of its edges slightly chipped and top chiseled; otherwise in good condition. Lettering on east and west sides.

Monument No. 18

This monument is located at Rossville, S. I., on lands of Mrs. Harriet Clark, on the southerly side line of the Shore road, 22.6 feet easterly from the line fence between lands of Mrs. Clark and V. C. F. Disoway and about 70 feet west of St. Luke's avenue. At the monument the following bearings and distances were taken: N 80° E, 21.0 feet to a cherry tree; N 30° W, 54.2 feet to the southwest corner of a frame storehouse, and S 20° W, 65.3 feet to the northeast corner of Disoway's house. This monument has its northeast and southeast corners slightly chipped and top chiseled; otherwise in good condition. Lettering on east and west sides.

Monument No. 19

This monument is located about one-quarter mile east of the Rossville, S. I., post-office, on land of Franklin Post, on the northerly slope from Fresh Kills road, about 10 feet from line between upland and meadows, surrounded with sassafras bushes and poison ivy. At the monument the following bearing and distance was taken: S 45° W, 9.0 feet to a small sassafras tree. This monument leans a trifle to the north and has its northerly and westerly corners slightly chipped and top chiseled; otherwise in good condition. Lettering on northeast and southwest sides.

This monument may be found by producing northerly the easterly side, or line, of the large frame dwelling with four large columns in the front, now occupied by Mrs. A. G. Decker, and from thence going easterly 46.0 feet to the monument.

Monument No. 20

This monument is located in the borough of Roosevelt, N. J., on land of the United States Metal Refining Co. It may be located as follows: Produce the center line of iron columns forming the southerly side of the smelter building and measure 79.0 feet westerly from the center of the corner iron column and from said point measure 33.5 feet southerly at right angles from the aforesaid line to the monument, which is encased in a wooden box, 10 inches by 12 inches, which stands just northeast of the northeasterly rail of the granulated slag track of the company, and about 2 feet below the top of ties. This monument, owing to the kind consideration of the company, has been carefully located and protected as above and in consequence is now readily found. It is in good condition. It should be raised to grade and set in concrete.

Monument No. 21

This monument is located in the borough of Roosevelt, N. J., on land of the Goldschmidt Detinning Co., 6 feet south of new north fence of said company. It may be located as follows: 87.0 feet east of the easterly side, or line, of the Goldschmidt Detinning Company's frame building produced northerly and 10 feet west of a small ditch dividing the meadow from the upland. This monument leans a trifle to the north and projects more than usual; otherwise in good condition.

Monument No. 22

This monument is located in the borough of Roosevelt, N. J., on land of the American Agricultural Chemical Co. (Liebig works). It may be located as follows: 499 feet southeasterly from monument No. 23, 10.7 feet easterly from the easterly side, or line, of the brick digester plant (main building, not extension) produced northerly and 21.3 feet north of the northerly side, or line, of the aforesaid building produced easterly. This monument has its corners and edges worn and chipped; otherwise in good condition.

Monument No. 23

This monument is located in the borough of Roosevelt, N. J., on upland of the American Agricultural Chemical Co. (Liebig works). At this monument the following bearings and distances were taken: N 50° W, 42.6 feet to a swamp oak tree stump standing alone; S 30° W, 95.5 feet to the northeast rail of a double track branch railroad, or spur, running into the Liebig works;

S 35° W, 175 paces to high brick stack of Liebig Co., and easterly about 10 feet to an open ditch, or drain. This monument has been covered by dredge fill. A section of steel rail 6 feet high marks the point. This monument leans a trifle to the south and has all four corners and two of its edges slightly chipped; otherwise in good condition. It should be raised to grade and set in concrete.

Monument No. 24

Quoting from the Engineers' joint report of October 8, 1913: " This monument is located in the borough of Roosevelt, N. J., on land of the Warner Chemical Company. It is 8.8 feet southwesterly from the northeasterly side, or line, produced and 7.4 feet southeasterly from the southeasterly side of the main brick building of the Company . . . It is 323.27 feet northerly from U. S. monument 'N' at the foot of Rahway avenue and at an angle of 68° 57' 54" in the northeast quadrant from monument 'N' and Melvin's chimney on Staten Island. It is . . . about 2 feet below the outside surface of the ground. It is protected by one length of 8-inch vitrified pipe, covered with a 6-inch yellow pine cover, over which there is about 6 inches of earth. It has its westerly corner slightly chipped, otherwise in good condition." This monument is entirely surrounded by buildings and should be located at a more suitable point.

Monument No. 25

This monument is located at Linoleumville, S. I., in edge of meadow and upland of American Linoleum Manufacturing Co., about 600 feet north of the northerly fence of the present plant of the company, which runs about N 60° E and S 60° W. At the monument the following bearings and distances were taken: N 35° E, 12.8 feet to a small oak stump; S 80° E, 25.8 feet to an oak stump; S 50° W, 25.8 feet to an oak stump 3 feet high, and N 40° W, about 5.0 feet to the line between upland and meadow. The monument has its corners and three of its edges chipped; otherwise in good condition.

Monument No. 26

This monument is located in the borough of Roosevelt, N. J., on land of Geo. F. Gantz. At the monument the following bearings and distances were taken: N 25° E, 60.5 feet to the remains

of an old stone wall (the highest stack is on this same range);
S 55° E, 126 feet to a small ditch about at the edge of the meadow
(the stack of the Williams & Clark Fertilizing Works is on this
same range); and S 40° W, 132.2 feet to the nearest rail of three-
track branch railroad running to the Williams & Clark and other
plants, and an additional distance of about 250 feet to a pin oak
tree on this range, between poles Nos. 21 and 22 along railroad;
S 65° W to northeast corner apple orchard and house belonging
to Henry Bunce 500 feet back of Rahway avenue. This monument has its easterly corners and easterly edge slightly chipped
and westerly edge broken and projects more than usual; otherwise
in good condition.

Monument No. 27

This monument is located at Linoleumville, S. I., on land of
Peter Cannon, about 285 feet N 15° E from house now occupied
by Mrs. Murphy, on southerly side of Burning Hill road. At the
monument the following bearings and distances were taken:
S 80° E, 22 yards to an English poplar tree; S 45° E, 25 yards to
a twin English poplar tree; S 5° W, 38 yards to another English
poplar tree, and N 80° W, about 167 feet to monument No. 28.
Monument No. 27 is set in concrete and about one foot above
ground and has its corners and northerly edge slightly chipped:
otherwise in good condition.

Monument No. 28

This monument is located at Linoleumville, S. I., on land of
Peter Cannon. It is about 167 feet northeast of the center of
Burning Hill road, about 20 feet southeast of creek and about
167 feet west of monument No. 27. This monument has its
corners and edges slightly chipped; otherwise in good condition.
It is set in concrete and extends a foot above ground.

Monument No. 29

This monument is located at Linoleumville, S. I., in meadow
land now or formerly of Mr. Meyer. It is about 200 feet northeast of a large creek and about 50 feet east of the kill, 60 feet more
or less west of mud bank and 6 inches above meadow. This monument has its corners and easterly edge chipped; otherwise in good
condition.

Monument No. 30

This monument is located in the township of Linden, N. J., on land of the Montgomery Co., about 160 feet south of the Tremley lane and 15 feet north of a small ditch dividing the upland from the meadow. At the monument the following bearings and distances were taken: N 80° E, 45.2 feet to a small twin pear tree; S 20° E, 4.6 feet to a small thorn-apple tree, and N 20° W, 9 paces to a twin maple tree. This monument leans a trifle to the south and west; otherwise in good condition.

Monument No. 31

This monument is located in the township of Linden, N. J., on land of the Montgomery Co., 355.5 feet northeast of monument No. 30. At the monument the following bearings and distances were taken: N 60° E, 81.0 feet to the nearest rail of a single-track branch railroad, called Carteret branch of C. R. of N. J.; N 10° E, 53.0 feet to an angle in a ditch separating the upland from the meadow. This monument leans a trifle to the west and has its corners and easterly edge broken; otherwise in good condition.

Monument No. 32

This monument is located in the township of Linden, N. J., on land of the Montgomery Co., several hundred feet north of Tremley lane and on the north edge of woods near where it turns sharply to the west. At the monuments the following bearings and distances were taken: N 65° W, 38.0 feet to a small twin oak tree at the corner of the woods where it turns sharply to the west; N 25° E, 81.0 feet to a ditch 63 feet south of the angle in the woods (the highest stack is on this same range); S 50° W, 30.7 feet to an oak tree, and S 10° W, 26.7 feet to another oak tree standing on the edge of the woods. This monument leans a trifle to the southwest and has its easterly and westerly corners and northerly and southerly sides broken; otherwise in good condition.

This monument is on the edge of meadow, N 40° W, 800 feet, more or less, from monument No. 31; N 65° W from crematory stack on Richmond Hill, S. I.

Monument No. 33

This monument is located on Prall's island, about 800 feet south of northerly end and about 240 feet west of easterly side of island, also about 120 feet west of swamp hole about 25 feet in diameter, also 44 feet east of the range of a stack in Chelsea and a church spire in Linoleumville, the bearing of the range being about S 10° E. This monument was not found in consequence of this section of the island having been filled in.

Monument No. 34

This monument is located at Bloomfield, S. I., on meadow land now or formerly of J. S. Drake, about 900 feet north of the kill opposite Prall's island, about 1,000 feet east of the kill around the turn, or bend, from the above, and about 750 feet N 60° W from U. S. C. S. monument No. 24. This monument may be readily located by measuring 6.0 feet northeast of the range of a frame building with cupola and flag pole (church or school) and stack of crematory on the summit of Richmond, bearing S 55° E. (After careful search of more than a half day and following description carefully, could not find this monument.)

Monument No. 35

This monument is located at Bloomfield, S. I., on upland now or formerly of Mrs. Scudder, on the north side of the second point of upland south of the B. & O. R. R., about 20 feet south of the edge of the meadow. (It may be reached via Lambert avenue.)

At the monument the following bearings and distances were taken: N 65° E, 38.3 feet to a white oak tree standing near the edge of the meadow; S 30° W, about 500 feet to chimney of house occupied by Rooney Decker (house burned down and chimney gone May 15, 1919); N 80° W, 115.2 feet to monument No. 36, and N 20° W, 13.7 feet to a four-legged white oak standing near the edge of the meadow. This monument has its corners and edges slightly chipped; otherwise in good condition. It is set in concrete.

This monument may be readily located by measuring 124 feet westerly from the range of water-tank and stack of Procter & Gamble's soap plant.

Monument No. 36

This monument is located at Bloomfield, S. I., on upland now or formerly of Mrs. Scudder, on the north side of the second point of upland south of the B. & O. R. R., about 75 feet east of the point and 5 feet south of the edge of the meadow. At the monument the following bearings and distances were taken: S 80° E, 115.2 feet to monument No. 35; about due south, 12.5 feet to a three-legged black oak and 27.0 feet to a four-legged black oak, and N 80° W, 23.7 feet to a small white oak tree standing near the edge of the meadow. This monument has three of its edges slightly chipped; otherwise in good condition.

Monument No. 37

Quoting from the Engineers' joint report of October 8, 1913:

"This monument was located on Buckwheat island in Arthur Kill, but, from measurements made on September 11, 1913, we find that it has been removed from the erosion of that portion of the island on which it stood."

Monument No. 37-A

Quoting from the Engineers' joint report of October 8, 1913:

"This monument, granite, eight inches by eight inches by 3.7 feet long, lettered 'N. Y.–N. J.' on its northerly side and '37-A' on its southerly side, was located and set in concrete on Buckwheat island in Arthur Kill, on September 25, 1913, as follows: Seventy feet westerly from the position of monument No. 37 as defined on the 'Revised Map of Pierhead and Bulkhead Lines for both shores of Arthur Kill or Staten Island Sound, from Raritan Bay to Storys Flats, as recommended by the New York Harbor Line Board, April, 1911,' approved September 28, 1911, and at right angles with the line joining said monument No. 37 as above defined and monument No. 40.

"The position of monument No. 37 as adopted and used in the location of monument No. 37-A with reference to other monuments, was as follows: In line, or range, with monument No. 33 on Chelsea, or Pralls island and U. S. H. L. monument 'B' on Buckwheat island and distant 191.9 feet northerly from U. S. H. L. monument 'A,' also on Buckwheat island. This latter monument 'A' was found to be 0.39 foot east of the said line joining monuments No. 33 and 'B,' which variation from alignment is very likely due to the dredging of the channel of Arthur Kill, which approaches nearer the southerly end of the island than it does the northerly end and hence its disturbing effects are, greater at 'A' than at ' B.' "

This monument was found to be in good condition.

Monument No. 38

This monument is located at Linden, N. J., on upland of the Standard Oil Co. of N. J., about 40 feet west of line between upland and meadow. It is about 540 feet southwest of a single-track branch railroad running through Railroad avenue from Standard Oil plant to C. R. R. of N. J., Long Branch division. At the monument the following bearing was taken: N 30° W, about 325 feet to gas tank No. 2, the middle one of three standing alone. It is 180.4 feet east of wire fence on easterly right-of-way line of Public Service Fast Line to Perth Amboy and 278 feet east of picket fence east of gas tank No. 2. This monument has three of its corners broken and its westerly edge chipped and leans a trifle to the northwest; otherwise in good condition.

To reach monuments Nos. 38 and 39, enter gate at Railroad avenue, proceed easterly to Public Service Fast Line tracks. No. 38 is south of and No. 39 north of this intersection.

Monument No. 39

This monument is located at Linden, N. J., on upland of the Standard Oil Co. of N. J., about 5 feet north of line between upland and meadow and about 1,200 feet northeast of monument No. 38 and nearly opposite tank No. 421 of Standard Oil Co. of N. J. It may be readily located by measuring 5.8 feet easterly from the easterly side, or line, of brick paraffin pump-house produced and 141.0 feet southerly from the southerly side, or line, of the aforesaid building produced, 22.1 feet east of east gage line of northbound Public Service track and 19 feet west of fence on east side of right of way of said railway. This monument has its corners and edges chipped; otherwise in good condition. (See note under No. 38 for directions to reach this monument.)

Monument No. 40

This monument is located at Bay Way, N. J., on land of the Waclark Wire Co., or Realty Co., near the mouth of Morse's creek, 5 feet northwest of high-water line of Arthur kill, 100 paces southwest of high fence of Waclark Co. At the monument the following bearing and distances were taken: N 65° E, about 390 feet to stack of the Waclark Wire Co., and 65 paces east of west curb of South Front street. This monument has its northeast corner and northerly and southerly edges slightly chipped; otherwise in good condition. It is set in concrete.

Monument No. 41

This monument is located at Holland Hook, S. I., on meadow land of the Staten Island Rapid Transit Railway Co., about 40.0 feet southwest of southwest foot of slope of the B. & O. R. R. embankment, also about 250 feet northeast of Old Place creek and about 350.0 feet southeast of southeasterly abutment of the B. & O. R. R. bridge. This monument was found to be in good condition in every respect.

Monument No. 42

Quoting from Mr. Hopper's report of 1916: "This monument is located at Holland Hook, S. I., on meadow land of J. I. Housman and John Croak, about 300 yards northeast of B. & O. R. R., about 350 yards southeast of the kill and about 120 feet southwest of a creek 10 feet wide.

"At the monument the following bearings were taken: S 85° E to Procter & Gamble's stack; S 40° E to stack of crematory on the summit of Richmond; S 55° W to Standard Oil Co. stack, and N 15° E to twin church spires in Elizabethport.

"This monument was found to be in good condition in every respect, and may be readily located by measuring 6 feet northeasterly from the range bearing about S 65° E of Procter & Gamble's steel water-tank and the next steel water-tank southeast thereof; also by measuring northwesterly 17 feet from the range bearing about N 30° E of flag-pole on Recreation pier and southeasterly spire of twin spires of church in Elizabethport."

As meadow has been filled and some of the range points are gone, we were unable to find this monument.

Monument No. 43

This monument is located at Holland Hook, S. I., on meadow land of John J. Dooley, about opposite Elizabeth Yacht Club house and about 250 feet south of kill. This monument may be readily located by measuring 28 feet easterly from the range of flag-staff on Recreation pier and church with cross on its spire, (both in Elizabethport), bearing N 35° W, also by measuring S 65° W, 177.2 feet from monument No. 44. Monuments Nos. 43 and 44 are both in range with the center of the third bay from the southeasterly end of steel lumber shed and flag-staff on Bohemian Hotel at the ferry to Elizabethport. This monument was found to be in good condition in every respect.

Monument No. 44

This monument is located at Holland Hook, S. I., on meadow land of John J. Dooley, about 100 feet south of kill. At the monument the following bearing and distance were taken: S 65° W, 177.2 feet to monument No. 43, which is in the range of monument No. 44 and centre of third bay from the southeasterly end of steel lumber shed and flag-staff on Bohemian Hotel at the ferry to Elizabethport. This monument was found to be in good condition in every respect. (It is difficult to locate except by measure from monument No. 43, as it is level with the meadow.)

Monument No. 45

This monument is located at Holland Hook, S. I., on meadow land of John J. Dooley, about 140 feet north of creek bridge and about 175 feet west of a branch railroad track from Procter & Gamble's plant to the dock. This monument may be readily located by measuring 27 feet southerly from the range, bearing N 70° W of the southerly corner of the buff brick hotel and southerly side of gas holder, both in Elizabethport, also by measuring 17.5 feet north of the range bearing N 85° W of black, or steel, stack in Elizabethport and the John Stevenson Electrical Sight Clock Tower and Cage (the case has been removed, leaving a portion of the tower which supports the tank); also by measuring 12 feet westerly from the range of Procter & Gamble's stack (second layout) and steel water-tank.

This monument could not be found. The ranges are gone. Probably it has been covered over by the Procter & Gamble Co.

Monument No. 46

This monument is located at Port Richmond, S. I., on land of John J. and Mary A. Worth, No. 2233 Richmond terrace. This monument may be readily located by measuring 24.3 feet easterly from a line fence and 43.0 feet southerly from the face of a stone bulkhead of the kill in front of this land. This monument is under a privet hedge on east side of cement walk, now covered with about twelve inches of soil, and is in good condition.

Monument No. 47

This monument is located at Port Richmond, S. I., on land of the Standard Oil Co. of N. Y., at No. 2201 Richmond terrace. This monument may be readily located by measuring 3.75 feet northeast from the line fence between lands of the above company and J. E. Donovan, also by measuring 15.7 feet southwesterly from the concrete water white oil building and 28.1 feet southeasterly from the northwesterly side, or line, of the brick gasoline building produced. This monument is now covered with about three feet of filling and has a plank covering on terra-cotta pipe, the plank about 12 inches below surface. It has its northerly corners chipped; otherwise in good condition.

Monuments Nos. 48 and 49

These monuments are both located at Bergen Point, Bayonne, N. J., on land of The Safety Insulated Wire & Cable Co. (No. 48 is under barrels. Superintendent assured us it had not been disturbed.) Quoting from the Engineers' joint report of October 8, 1913:

" No. 48 is within the frame building No. 28. It is 6.9 feet northwesterly from the southeasterly side and 21.5 feet southwesterly from the northeasterly side of said building No. 28. It is about 2.5 feet below the surface of the ground and is to be protected by the company, by a concrete casing 20 inches square on the inside and 32 inches square on the outside at the top and covered with a suitable cover. It has its corners and edges chipped and leans slightly to the northwest, otherwise in good condition."

This monument is covered by a huge pile of barrels. The superintendent assured us it is in good condition.

Again quoting from the Engineers' joint report of October 8, 1913:

" No 49 is 101.65 feet southeasterly from the southeasterly side of the brick building No. 1 and 109.8 feet northeasterly from the southwesterly side, or line, of said brick building No. 1 produced southeasterly. It is about 1.5 feet below the surface of the ground and is to be protected by the company, by a concrete casing 20 inches square on the inside and 32 inches square on the outside at the top and covered with a suitable cover."

It is in good condition.

Monument No. 50

This monument is located at Bergen Point, Bayonne, N. J., on land now or formerly of Rufus Story. This monument is about 2.7 feet south of the south line of First street, 14.7 feet south of south curb line and about 47.7 feet west of the west line of Lord avenue produced. This monument has its corners and easterly and westerly edges slightly chipped; otherwise in good condition.

Monument No. 51

This monument is located at West New Brighton, S. I., on land of C. W. Hunt, in rear of frame building No. 1589 Richmond terrace, now used as rope department by the owner. It is about 4.5 feet below the present surface of the ground and is protected by brickwork with an iron cover. This monument may be readily located by measuring 15.95 feet northerly from the northerly side and 29.2 feet easterly from the westerly side produced of the above building, also by measuring 39.1 feet southerly from the southerly rail of the east-bound track of the Staten Island Rapid Transit railway, also by measuring 114.5 feet easterly from the westerly line fence of the C. W. Hunt property. This monument has its northeasterly edge slightly chipped; otherwise in good condition. It should be raised to grade and set in concrete.

Monument No. 52

This monument is located at Bergen Point, Bayonne, N. J., on land of the Gulf Refining Co., about 4 feet below the surface of the ground and is protected by the company by a concrete casing 20 inches square on the inside and 32 inches square on the outside at the top and covered with a 3-inch yellow pine cover. It may be readily located by measuring 44.46 feet easterly from the easterly side of oil tank No. 206, and 48.15 feet southerly from the southerly side of oil tank No. 211. This monument leans a trifle to the east and has its corners and northerly and southerly edges slightly chipped; otherwise in good condition.

Monument No. 53

This monument is located at Livingston, S. I., on land (right of way) of the Staten Island Rapid Transit Railway Company.

Quoting from the Engineers' joint report of October 8, 1913:

"We find from the best information now obtainable that on or about August or September, 1895, this monument was removed from its original position by Mr. A. B. Proal, at that time engineer for the Richmond Light and Railroad Company, and reset about 23 or 26 feet northerly in line, or range, with its original position and the Tide Water Oil Company's chimney on Constables Hook, Bayonne, N. J., by means of a survey made for the purpose by Mr. William MacDonald; also that its position as thus moved and reset was subsequently recorded in Mr. John C. Payne's field book as 19.2 feet northerly from the northeasterly corner of the brick pump room of the power plant of the Richmond Light and Railroad Company, at Livingston, S. I., and 0.5 foot westerly from the easterly side, or line, of said brick pump room produced northerly.

"This monument was found to be leaning badly to the west. Hence, on September 12, 1913, it was carefully reset in concrete in its recorded position as above."

This monument is between the two main tracks of the Staten

Island Rapid Transit Railway and its top is about level with the top of ties of west-bound track. It has its corners and edges worn; otherwise in good condition.

Monument No. 54

This monument is located at New Brighton, S. I., on land of Sailors' Snug Harbor. This monument is about on line with the fronts of houses A, B, C, D and E, and in front of the Randall Memorial church. It is also 45.0 feet southeast of the southeast side of the marble fountain. This monument leans a trifle to the south and west and has three of its corners slightly chipped; otherwise in good condition.

Monument No. 55

This monument is at New Brighton, S. I. It is in the roadway of Richmond terrace, 6.5 feet south of north curb and about 165 east of east line of Westervelt avenue and within the west-bound trolley track. It is two inches below the surface of the roadway and is protected by a hinged iron cap. It may be readily located by measuring 7.15 feet west of the easterly side, or line, produced of the main frame building occupied by Richmond Council No. 351, Knights of Columbus (street No. 404), and by measuring 90.55 feet easterly from the southeast corner of brick building No. 421 Richmond terrace. This monument has its top corners and edges worn; otherwise in good condition.

Monument No. 56

This monument is located at St. George, S. I., on land of the B. & O. R. R. Co., at westerly end of yard and near base of steep slope to Richmond terrace. Quoting from the N. Y. State Engineer and Surveyor's report of 1900: "Monument No. 56 is at St. George, Staten Island, on the property of the B. & O. railroad at west end of yard, four feet from foot of terrace, 94 feet east of culvert, 1 foot from rail and 105 feet east of east end of wall at foot of terrace. This monument has its north and south sides broken."

As monument No. 56-A takes the place of this monument, no attempt was made to find monument No. 56.

Monument No. 56-A

Quoting from the Engineers' joint report of October 8, 1913:

" Mr. John C. Payne, secretary and engineer of the Riparian Commission of New Jersey, informs us that on September 3, 1904, this monument (granite) was carefully set in line, or range, with monument No. 56 and the Crude Oil Company's chimney on Constables Hook, Bayonne, N. J., and that its location was near the northerly side of Jay street, about opposite the junction of Stuyvesant place at St. George, S. I.

" In constructing the retaining wall along the northerly side of Jay street, the borough authorities, after first carefully noting the position of the above monument, removed it. After the retaining wall was completed and on or about September 25, 1912, the said borough authorities carefully set or caused to be set a brass bolt in the northerly concrete sidewalk of Jay street to mark the identical position of the former stone monument, which was removed as above.

" The present monument, No. 56-A, is therefore a brass bolt set in the northerly concrete sidewalk of Jay street about opposite the junction of Stuyvesant place, at St. George, S. I. It is 0.485 foot southerly from a drill hole and cut in the top of the coping of the retaining wall of Jay street."

This brass bolt is at the second lamp-post east of the west end of retaining wall; also it is 70.2 feet easterly from a drill hole and cut in the northerly curb of Richmond terrace. (Jay street is now Richmond terrace.)

This brass bolt was found to be in good condition.

MONUMENTS ON LANDS FROM HUDSON RIVER TO DELAWARE RIVER

A detailed description follows of the location and condition of the monuments marking that part of the New York and New Jersey boundary line extending from the Hudson river to the Delaware river.

Eastern Terminal Monument

The monument at the eastern terminus of the line is a large block of trap-rock, seven feet six inches long, three feet two inches high and about four feet thick. It is located at the foot of the Palisades, about six inches above storm tides of the Hudson river and about sixty feet west of the present shore line of the river. It is marked with a groove upon its perpendicular eastern face for its full heighth at a distance of two feet south from its northerly end and is further marked with the words " Latitude 41 degrees north " and on the north side of the groove the words " New York " and on the south side thereof the words " New Jersey," which last four quoted words are now covered by the cement masonry to hold in position a large pole, about eight inches in diameter and about thirty feet high, located about two inches easterly from the easterly face of the stone. Near the top of the pole and securely fastened thereto, there is a sign board about eighteen inchs wide by about five feet long, upon the easterly side of which appears the words, "State Line" and the letters, " N. J." and " N. Y." It lies 313.21 feet south 18° 44' west from the U. S. Coast Survey station Duer, and, from the determinations of that survey, it is in latitude 40° 59' 48.17" north, longitude 73° 54' 11" west from Greenwich.

Eastern Witness, or Reference, Monument

This monument is 488 feet from the terminal monument on the bank of the Hudson river and is 350 feet above tide level. It stands opposite a point on the New York Central railroad about midway between Dobbs Ferry and Hastings and the boundary line, if extended across the Hudson, would cross the railroad near the tall, old chimney south of Hastings. By clearing away bushes in its immediate vicinity the monument would be in plain sight of the east bank of the Hudson from near Ossining almost to Yonkers. The monument is in one piece, eleven and a half feet long with a cross-section of somewhat over one by two feet. It weighs nearly three tons and is set four feet down in an accurately cut hole in the rock and fastened with cement mortar and is further supported for a foot and a half more by building stones and hydraulic mortar around it. The remaining portion (six feet) is hammered, dressed, polished and marked on two of its sides.

The north side is marked:
NEW YORK

BOUNDARY
MONUMENT

1882

HENRY R. PIERSON,
CHAUNCEY M. DEPEW,
ELIAS W. LEAVENWORTH,
COMMISSIONERS.

488 FEET WEST
FROM STATION ROCK
H. W. CLARKE
SURVEYOR.

The south side is marked:
NEW JERSEY

BOUNDARY
MONUMENT

1882

ABRAHAM BROWNING,
THOMAS N. McCARTER,
GEORGE H. COOK,
COMMISSIONERS.

488 FEET WEST
FROM STATION ROCK
E. A. BOWSER
SURVEYOR.

This monument stands on the property of the Palisades Inter-State Park and aside from having all of its corners chipped it is in excellent condition. There is a vertical groove cut on the east and west ends of base of monument.

Monument No. 1, or Road Monument No. 1 between the Eastern Witness, or Reference, Monument and Milestone I

This monument is situated on the west side of the boulevard which leads from Alpine, N. J., to the village of Palisades, N. Y. It is at the east foot of the hill which rises about 15 degrees to its summit, distant about 100 paces, and is about 2,728 feet east of mile stone I. At the monument the following bearing and distance were taken: S 51° E, 5 feet 1½ inches to a whitewood tree. Soil sandy. Disk set in lime obtained from an old house. This monument was found to be in good condition except that all four corners were slightly chipped.

Monument No. 2, or Milestone I

This monument consists of two stones, the old original brown-stone monument and the new granite monument, set east of and adjacent to the old one. It is situated on the west slope of a ridge between the boulevard and the road from Closter to Palisades. It is about 971 feet east of the monument No. 3 and stands on land belonging to Jacob S. Moore, 23 paces north of a rail fence running up over the ridge.   At the monument the following bearings and distances were taken: S 85° W, 18 paces to a large oak, and S 72° W, 14 paces to another.   Soil sandy.   Disk set in leaves.   This monument was found to be in good condition except that the northerly and southerly corners were chipped.

Monument No. 3, or Road Monument No. 1 between Milestones I and II

This monument is situated on the west side of the road which leads from Closter to Palisades and is about 971 feet west of milestone I.   At the monument the following bearing and distance were taken: N 50° E, 30 paces to a large black oak which stands on the west side of the road and just in front of David Munson's house.   Soil sandy for a depth of three feet, when solid rock was encountered.   Disk set in wood ashes.   This monument was found to be in good condition except that the corners were slightly chipped.

Monument No. 4, or Road Monument No. 2 between Milestones I and II

This monument is situated on the west side of a road leading from Closter to Sparkill and is about 2,992 feet west of milestone I and on ground sloping gently westward.   At the monument the following bearing was taken: north by west 5 feet to a shade maple. Soil sandy.   Disk set in wood ashes.   This monument leans very slightly to the north and has its corners chipped; otherwise it is in good condition.   There is a sign on a maple tree near this stone marked } "State Line N. Y. N. J."

Monument No. 5, or Milestone II

This monument is situated in the Tappan timber swamp, on land of Joseph Leival, and is about 2,304 feet west of monument No. 4, on ground which is low all around and at times covered with water. At the monument the following bearing and distance were taken: About N 45° W, 25 paces to a marble monument four inches square the corner of McCreary in Leival's line.   Soil clayey.   Disk set in wood ashes.   This monument (both stones) was found to be in good condition in every respect. The granite stone is one foot square.

Monument No. 6, or Railroad Monument No. 1 between Milestones II and III, on Northern Railroad of New Jersey

This monument is level with the surface of the ballast of the road-bed of the Northern Railroad of New Jersey, 2 feet 11 inches west of the westerly rail and 104 feet 4 inches southerly from the center (at the westerly end) of the center pier of the railroad bridge over small stream, Sparkill; it is also 1,501 feet west of milestone II and 327 paces southeast of Tappan station. This stone is one foot square and in good condition. Westerly corners slightly chipped. Disk set in ashes from the railroad and 15 inches below the bottom of the stones.

Monument No. 7, or Road Monument No. 1 between Milestones II and III

This monument is situated on the west side of the road leading from Norwood to Tappan, on land of William Rogers, and is about 2,904 feet west of milestone II and 130 paces north of the highway bridge over Sparkill creek. At the monument the following bearing and distance were taken: N 30° E, 75 paces to the southwest corner of a barn which stands just east of the road. Soil sandy. Disk set in wood ashes. This monument was found to be in good condition except that the corners were slightly chipped.

Monument No. 8, or Road Monument No. 2 between Milestones II and III

This monument is situated on the east side of the road leading from Schraelenburgh to Tappan, about 4,400 feet west of milestone II and on ground rising gently northwest. At the monument the following bearings and distances were taken: North 76 paces to the southeast corner of the Dutch Methodist church standing just east of the road; about north 48 paces to the front entrance of Wilson's yard, also about 10 feet northerly to a cedar tree standing on the same side of the road. Soil sandy and somewhat stony. Disk set in wood ashes. This monument has its corners and edges badly chipped; otherwise it was found to be in good condition.

Monument No. 9, or Railroad Monument No. 2 (West Shore Railroad) between Milestones II and III

This monument is situated between the two westerly tracks of the West Shore railroad and 45 paces S 20° W of Andre avenue. At the monument the following bearing and distance were taken: N 50° W, 67 paces to monument No. 10. Soil sandy. Disk set in wood ashes. This monument is level with the subgrade, 6 inches square, and has all four corners and edges badly chipped; otherwise it was found to be in good condition.

Monument No. 10, or Road Monument No. 3 (Andre Avenue) between Milestones II and III

This monument is situated on the east side of Andre avenue in front of land of Mrs. Ellen Watson. It is about 5,051 feet west of milestone II and is 13 paces south of the westerly fence of the avenue and on the east slope of the hill. At the monument the following bearing and distance were taken: N 10° W, 30 paces to the middle house of three which stand just north of the road, or avenue. Soil sandy. Disk set in wood ashes. This monument has its easterly corner slightly chipped; otherwise in good condition.

Monument No. 11, or Milestone III

This monument is situated on the east slope of Andre hill, property of Mrs. Ellen Watson, about 584 feet from the summit or about half-way up. At the monument the following bearings and distances were taken: S 40° E, 36 paces to the southwest corner of a little white house, and S 70° W, 18 paces to an old well. Soil of fine gravel. Disk set in wood ashes. This monument (both stones) was found to be in good condition except that the new stone has its southeasterly corner chipped and the old stone leans a trifle to the south.

Monument No. 12, or Road Monument No. 1 between Milestones III and IV

This monument is situated on the top of Andre hill, about 584 feet west of milestone III, on land of Mrs. Ellen Watson. At the monument the following bearings and distances were taken: South 19 paces to the northeast corner of Mrs. Watson's house and N 15°

E, 52 paces to the Andre monument. Disk set in wood ashes. This monument was slightly chipped on its four corners; otherwise it was found to be in good condition.

Monument No. 13, or Road Monument No. 2 between Milestone III and IV

This monument is situated on the north side of the road leading from Tappan to Rivervale, about 1,825 feet west of milestone III and on ground comparatively level. It is 24 paces northwesterly of a small culvert which is the beginning point of two road districts and 20 paces easterly of the entrance into Joseph Mack's barn-yard. At the monument the following bearing was taken: N 51° W to Mack's house. Soil sandy. Disk set in cinders. This monument was found to be in good condition in every respect.

Monument No. 14, or Milestone IV

This monument is on land of Chas. Smith. At the monument the following bearings and distances were taken: N 20° E, 3 feet to a cedar stump; N 10° W, 15 feet to clump of white birches, and N 10° W, 22 paces to a large whitewood tree standing in line with the stone wall marking the southeasterly side of a woods road leading northeasterly from the main road, which runs from West Norwood, N. J., to Orangeburg, N. Y., and distant about 800 feet northeasterly therefrom. The said woods road leaves the main road about 230 feet southeasterly of the old Duryea house. Soil sandy. Disk set in sand and monument in cement. This monument (both stones) was found to be in good condition in every respect. This territory is starting to grow up in brush.

Monument No. 15, or Road Monument No. 1 between Milestones IV and V

This monument is situated on the east side of a road leading from Rivervale to Orangeburg and running about north and south. It is about 2,419 feet west of milestone IV, eight paces west of fence on east side of road and 12 paces south of a small terra-cotta pipe culvert and on level ground. At the monument the following bearing and distance were taken: N 10° E, 111 paces to an old stone house owned by Oscar Dreisbach. Soil somewhat gravelly. Disk set in light-colored sand. This monument is covered by 6 inches of road material and was not found.

Monument No. 16, or Milestone V

This monument is situated in the woods about 50 paces west of a large cultivated field now owned by Broadacre Dairy Farms and south of cleared waste field containing many cedars, opposite a point about 200 feet northerly from the southwest corner of said cleared lot and about half-way up the slope of a slight rise. It is also 5,261.27 feet west of milestone IV and 40 paces west of an old woods road running north and south. It is 140 paces south of the junction of this road with a road across meadow, leading westerly from house formerly occupied by James Cassidy and now owned by Wilder, Erwen and Patterson. At the monument the following bearing and distance were taken: S 5° E, 4 feet to an oak stump. Soil sandy. Disk set in leaves. This monument (both stones) was found to be in good condition except that the southeast corner of granite stone was slightly chipped.

Monument No. 17, or Road Monument No. 1 between Milestones V and VI

This monument, reset in concrete July 9, 1913, is situated on the west side of a road running northeasterly and southwesterly and leading to Blue Hill. It is also about 4,298 feet west of milestone V and 68 paces south of the bridge. At the monument the following bearing and distance were taken: N 80° E, 138 paces to a large locust tree standing in Mr. Priest's front yard. Soil wet and gravelly. Disk set in wood ashes. This monument has three of its corners slightly chipped; otherwise in good condition.

Monument No. 18, or Milestone VI

This monument is situated on the east slope of the hill, about 200 feet within the wood-lot, now partly cleared, belonging to Mr. Curtis (Doriskill farm). It is 5,225.18 feet west of milestone V. At the monument the following bearings and distances were taken: S 51° E, 7 feet to an elm or whitewood tree, and N 10° W, 4 paces to a maple. Soil sandy. Disk set in sand. This monument (both stones) was found to be in good condition in every respect.

Monument No. 19, or Road Monument No. 1 between Milestones
VI and VII

This monument is situated on the west side of a road running
S 75° E and N 75° W and is about 3,408 feet west of milestone
VI and 13 paces north of an eighteen-inch ash tree on the westerly
side of the road. At the monument the following bearings and
distances were taken: S 70° E, 12 paces to a maple which stands
between the east edge of the road and the stone fence; N 30° E,
55 paces to a thorny locust tree which stands in the front yard of
W. Comes, and 7 paces south of concrete road post. Soil sandy.
Disk set in wood ashes. This monument was found to be in good
condition, except that two of its corners were slightly chipped.

Monument No. 20, or Milestone VII

This monument is situated very near the summit of the first
ridge west of the road in an open field on the land of C. W.
Dutcher and is about 5,241 feet west of milstone VI and 1,833
feet west of monument No. 19; also 13 paces easterly from the
westerly stone wall fence of the field; also the line crosses the
stone wall fence between the open field and the woods at a distance
of about 130 paces northerly along the same from the lane. Soil
sandy. Disk set in sand and monument in cement. The old
brownstone monument is broken and split. The new monument
was found to be in good condition, but loose, and should be reset
in concrete.

Monument No. 21, or Railroad Monument No. 1 between Mile-
stones VII and VIII, on the New Jersey and New York
Railroad

This monument, 12 inches by 12 inches, is situated on the west
side of the track of the New Jersey and New York railroad, close
in by the end of the ties, about 1,640 feet west of milestone VII,
about 880 paces along the track from the southeast corner of the
depot at Pearl River, and in a slight cut. This monument was
found to be in good condition, except that its westerly corners
were slightly chipped. There is a sign-post on the west side of the
cut, 3 inches by 12 inches by 10 feet high, marked " N. Y. N. J."

Monument No. 22, or Road Monument No. 1 between Milestones VII and VIII

This monument is situated on the west side of a road leading from Mount Vale to Pearl River and running about north and south, at the east foot of the hill, about 2,063 feet west of milestone VII and 385 feet west of monument No. 21, also about six feet north of the corner of a line fence running about N 51° W up over the hill. Soil sandy. Disk set in wood ashes. This monument was found to be in good condition, except that the easterly corners were slightly chipped.

Monument No. 23, or Milestone VIII

This monument is situated in an open field belonging to Mrs. Laura B. Hollis, on slightly rolling ground, about 5,251.25 feet west of milestone VII and about 25 paces west of the nearest point of the Pascack creek. At the monument the following bearing and distance were taken: Due west 15 paces to an apple tree. Soil sandy. Disk set in sand and the monument in cement. This monument, small one, is on a sloping hillside, leans badly to the southeast and should be reset in concrete. The old stone leans badly to the north.

Monument No. 24, or Road Monument No. 1 between Milestones VIII and IX

This monument, reset in concrete July 9, 1913, is situated on the east side of the road which runs north and south and is about 449 feet west of milestone VIII, about 10 feet north of a small stream on grounds sloping eastward and about 5 feet north of parapet wall of culvert. Soil very gravelly and the hole full of water. Disk set in wood ashes. This monument was found to be in good condition in every respect.

Monument No. 25, or Road Monument No. 2 between Milestones VIII and IX

This monument, moved October 10, 1910, and reset July 11, 1913, is situated on the north side of the road from Pearl River to Saddle River running nearly east and west, just at the top of a hill and about 2,320 feet west of milestone VIII. At the monument the following bearings and distances were taken: South 9 paces to a chestnut tree, now dead, on opposite side of the road, and due west about 6 feet to a white oak tree. Soil sandy. Disk set in sand. This monument was found to be in good condition except that its corners and edges were slightly chipped.

Monument No. 26, or Road Monument No. 3 between Milestones VIII and IX

This monument is situated on the west side of the road, about 3,492 feet west of milestone VIII and nearly opposite the point at which a road running north and south enters the main road on which the monument is set. The main road runs from Nanuet to Saddle River and the cross road to Pearl River. At the monument the following bearings and distances were taken: S 50° E, 22 paces to a small white oak standing just inside the fence and at the intersection of the roads, and N 51° W, about 400 feet to the house at top of hill. Soil somewhat gravelly. Disk set in wood ashes. This monument was found to be in good condition in every respect.

Monument No. 27, or Road Monument No. 4 between Milestones VIII and IX

This monument is situated on the west side of the road which runs about north and south, leading from Hackensack to Spring Valley, and is about 109.5 feet east of milestone IX and on level ground. Soil sandy. Disk set in wood ashes. This monument has its southerly edge slightly chipped; otherwise in good condition in every respect.

Monument No. 28, or Milestone IX

This monument is situated in a young orchard on the property of Mr. Post and on level ground. At the monument the following bearing and distance were taken: N 51° W, 115 paces to Post's house. Soil sandy. Disk set in sand and monument in cement. This monument (both stones) was found to be in good condition except that the old stone leans easterly and southerly. As this ground is regularly plowed, this monument should be set in concrete.

Monument No. 29, or Road Monument No. 1 between Milestones IX and X

This monument is situated on the west side of the road running about north and south, and is about 318 feet east of milestone X and on level ground. At the monument the following bearings and distances were taken: S 20° W, 55 paces to the northeast corner of John Foxlee's house, and N 32° E, 80 paces to the northeast corner of J. A. Christopher's house. Soil sandy. Disk set in wood ashes. This monument has its northerly edge slightly chipped; otherwise in good condition.

Monument No. 30, or Milestone X

This monument is situated about one foot north of a line fence between lands of John Foxlee and J. A. Christopher, just a little on the westerly slope of a slight rise and is about 5,286.6 feet west of milestone IX. At the monument the following bearings and distances were taken: S. 26° E, 135 paces to the northwest corner of Foxlee's house, and S. 78° W, 92 paces to a chestnut tree. Soil sandy. Disk set in sand and monument in cement. This monument (both stones) was found to be in good condition, except that the northerly corners of granite monument were slightly chipped.

Monument No. 31, or Road Monument No. 1 between Milestones X and XI

This monument is situated on the east side of a road leading from Saddle River and running S 20° W and N 20° E. It is about 4,050 feet west of milestone X and on ground sloping gently westward toward Saddle River. At the monument the following bearing and distance were taken: N 11° W, 37 paces to the southeast corner of foundation of Michael Connolly's barn, recently burned. Soil gravelly and sandy and hole very wet. Disk set in wood ashes. This monument was found to be in good condition in every respect.

Monument No. 32, or Milestone XI

This monument is situated in a rock pile in an open lot belonging to M. T. Connolly and is 5,267.6 feet west of milestone X and about 1,217 feet west of monument No. 31, on ground ascending westward, 11 paces from the wire fence separating apple orchard from open lot and 45 paces easterly from a small shed. Soil sandy and somewhat stony. Disk set in sand. This monument, small one, was found to be in good condition. The old stone leans southerly. Brush is starting to grow around these monuments and obscures them when foliage is out.

Monument No. 33, or Road Monument No. 1 between Milestones XI and XII

This monument is situated on the north side of the road which runs N 75° W and S 75° E, opposite an apple tree standing on the south side of the road. It is about 2,258 feet west of milestone XI and on the west slope of a hill, near the top. At the monument the following bearings and distances were taken: N 75° W, 20 feet to a large chestnut, now dead, standing on the north side of the road, and S 84° E, 32 paces (passing a wall at 7 paces) to the southwest corner of Morris Solperstein's house, which stands north of the road and about at the top of the hill. Soil sandy. Disk set in wood ashes. This monument was found to be in good condition, except that both easterly corners were slightly chipped.

Monument No. 34, or Road Monument No. 2 between Milestones XI and XII

This monument is situated on the east side of a road which runs northeast and southwest and is 3,585 feet west of milestone XI and about at the foot of the hill adjoining lands of Margaret DeBarry. At the monument the following bearing and distances were taken: N 25° E, 25 paces to a buttonwood tree, which stands near a bridge and on the east bank of the stream; also 20 feet to the center of said stream. Soil sandy. Disk set in wood ashes. This monument was found to be in good condition except that three of its corners were slightly chipped.

Monument No. 35, or Road Monument No. 3 between Milestones XI and XII

This monument is situated on the west side of the River Valley road, which runs about north and south, and is about 4,083 feet west of milestone XI. At the monument the following bearing and distances were taken: S 10° E, 31 paces along the road to a barn on the east side; 8 paces north of terra cotta pipe culvert set in concrete. Soil sandy. Disk set in wood ashes. This monument was found to be in good condition except that the southeast corner was slightly chipped.

Monument No. 36, or Milestone XII

This monument is situated in the wood-lot (about 60 feet west of easterly edge) belonging to Cornelius Snyder at the west foot of the hill and is about 25 paces N 51° W from the corner of the line fence which ends at the woods and 17 paces south of fence running east and west through the woods. At the monument the following bearing and distance were taken: North 16 paces to a hickory tree standing near a fence running east and west. Soil sandy. Disk set in sand. This monument (both stones) was found to be in good condition in every respect, except old stones broken on New Jersey face.

Monument No. 37, or Road Monument No. 1 between Milestones XII and XIII

This monument is situated at the top of a hill and on the east side of a road running north and south from Saddle River to Tallmans and is about 3,304 feet west of milestone XII. At the monument the following bearing and distances were taken: S 53° E, 12 paces to the northwest corner of H. Ackerson's house. Soil sandy. Disk set in wood ashes. This monument was found to be in good condition, except that the southeast corner was slightly chipped.

Monument No. 38, or Milestone XIII

This monument is situated on the west slope of a hill in an open field, the property of Louis H. Doremus, and 5,297 feet west of milestone XII. It is 16 paces east of west fence, 20 paces north of south fence and 50 paces east of Doremus' lane. Soil sandy. Disk set in sand and monument in cement. This monument, small one, has its southerly corners rounded; otherwise in good condition in every respect. Old, or large, stone leans slightly to the south.

Monument No. 39, or Road Monument No. 1 between Milestones
XIII and XIV

This monument is situated at the east side of a road running
N 40° E and S 40° W and is about 673 feet west of milestone
XIII. At the monument the following bearing and distance
were taken: S 18° E, 15 paces to the northwest corner of Louis
H. Doremus' house. Soil coarse gravel. Disk set in wood ashes.
This monument leans westerly and has its westerly corners chipped.

Monument No. 40, or Road Monument No. 2 between Milestone
XIII and XIV

This monument is situated at about the east foot of a hill and
on the west side of a road which runs N 20° E and S 20° W and
is about 2,782 feet west of milestone XIII. At the monument the
following bearing and distance were taken: About N 37° E, 30
paces to the southwest corner Geo. Dunlop's house. This monument is in the westerly prolongation of a line fence on the easterly
side of the road between land of G. W. Suterland and Geo. Dunlop. Soil sandy. Disk set in wood ashes. This monument has its
corners chipped; otherwise in good condition in every respect.

Monument No. 41, or Milestone No. XIV

This monument is situated near the edge of the woods on the
west slope of the hill and is 5,298 feet west of milestone XIII.
At the monument the following bearing and distances were taken:
S 42° E, 9 paces to a white oak stump; also about one-half mile
west of W. H. Way's house and at the corner of Melvin Brown's
land and Christie's and Geo. and A. Fox's land, Brown's fence
running westerly and Fox's running northerly. Also about 144
paces north of tower No. 149 of power line. Soil sandy. Disk
set in sand. This monument (both stones) was found to be in
good condition in every respect.

Monument No. 42, or Road Monument No. 1 between Milestones
XIV and XV

This monument is situated on the south side of the private road,
or lane, leading up to the " Foxwood Inn " and is 3,343 feet west
of milestone XIV and on level ground, outside of wire fence. At
the monument the following bearing and distances were taken:

N 85° W, 5 paces to a shade maple. Soil sandy. Disk set in sand.
Monument, being broken in the middle, was set in cement. This
monument was found to be in good condition except the northwest
corner and edges slightly chipped.

*Monument No. 43, or Road Monument No. 2 between Milestones
XIV and XV*

Quoting from the State Engineer's report of 1896:

" This monument is situated on the east side of the old Ramapo
post road, which runs about north and south, and is about 4,705
feet west of milestone ' XIV.' At the monument the following
bearing was taken: South 215 paces to the bridge over the Mah-
wah river; . . . Soil sandy. Disk set in ashes from the railroad. This monument is slightly chipped, otherwise it was found
to be in good condition."

It is about one foot below present surface of road; 48.55 feet
northeasterly from extreme corner of northerly wing wall; 110
feet northeasterly from extreme corner of southerly wing wall of
undergrade highway crossing, Erie railroad; 15.63 feet northwest from pole and 5.85 feet east of gage line of electric railway.
This monument is now covered by a concrete sidewalk, hence no
attempt was made to uncover it. New monument should be set
at a suitable point.

*Monument No. 44, or Railroad Monument No. 1 between Mile-
stones XIV and XV, on the New York, Lake Erie & Western
Railroad*

This monument, six inches by six inches square, of granite, is
situated between tracks numbered three and four of the Erie railroad at Suffern, N. Y., 6.8 feet east of easterly rail of track No.
4, 1.6 feet west of westerly rail of track No. 3 and 1.2 feet below
top of tie. It is about 124.25 feet westerly from monument No.
43 and 4,839 feet west of milestone XIV. It is 138.06 feet northerly from the easterly corner and 138.82 feet northerly from the
westerly corner of back wall of railroad bridge over highway.
This monument is under the road-bed and should be raised to
grade.

Monument No. 44-A, or Road Monument No. 3 between Milestones XIV and XV

This monument, 12 inches by 12 inches, marked by cross 8 inches long, thus $\dfrac{N \mid Y}{N \mid J}$ is situated on the easterly side of Ramapo avenue, 39.95 feet southeasterly from milestone XV. This monument was found to be in good condition.

Monument No. 45, or Milestone XV

This monument is situated near the center of Ramapo avenue, 39.95 feet from monument No. 44-A, and 5,280 feet west of milestone XIV. At the monument the following bearing and distance were taken: N 50° E, about 75 paces to Zabriskie's house. Soil sandy. Disk set in sand and monument in cement. This monument (both stones) is about one foot below the present bituminous macadam road surface, hence no attempt was made to uncover it.

Monument No. 46, or Road Monument No. 1 between Milestones XV and XVI

This monument is situated on the east side of the road which runs north and south along the east foot of the Ramapo mountains. At the monument the following bearing was taken: 2 feet west of small hickory standing opposite the end of a line fence. Soil sandy. Disk set in wood ashes. This monument has its corners chipped, the easterly ones badly; otherwise in good condition.

Monument No. 47, or Milestone XVI

This monument, granite, 12 inches square, is situated about half-way up the east slope of the Ramapo mountain and 10 feet east of a woods road, which leaves the main road nearly opposite monument No. 46. It is about 10,463 feet west of milestone XIV and about 200 feet westerly on line from the site of Peter Mann's cabin. At the monument the following bearing was taken: N 65° E to the spire of the Episcopal church in Suffern. Soil sandy to within about six inches of the full depth, when a rock ledge was encountered, on which was cut a cross in place of the disk. This monument was found to be in good condition in every respect.

Monument No. 48, or Milestone XVII

This monument is situated on the east slope of a ridge, about 26 paces southeast from the road leading past William DeGrott's house, and is 5,141.65 feet west of milestone XVI. At the monument the following bearings and distances were taken: About S 53° E, 12 feet to a red oak tree, and N 10° E, 8 paces to a black oak tree. Soil very rocky. Disk set in sand. This monument (both stones) was found to be in good condition except that new stone has its southeast corner badly chipped. The old stone is an undressed brownstone slab. The State line in this vicinity has recently been cut through.

Monument No. 49, or Milestone XVIII

This monument is situated in an open field belonging to the Pierson estate and about half-way down a gradual slope, which begins at the mountain on the southeast and slopes westward. It is 5,301.2 feet west of milestone XVII. At the monument the following bearing and distances were taken: N 85° E, 9 paces to a round top maple, also about 87 paces northerly to John Mann's house. Soil gravelly. Disk set in wood ashes. This monument (new stone) was found to be in good condition. Old stone is a small brownstone slab, not embedded, and leans northwardly.

Monument No. 50, or Road Monument No. 1 between Milestones XVIII and XIX

This monument is situated on the east side of a road running N 30° E and S 30° W, 156 paces southwest along said road from the center of a culvert. It is 197 paces from fork in road leading up the hill. The monument is about 8 paces north of a large boulder, which is about 12 feet long, 6 feet high and 10 feet wide. Soil sandy. Disk set in leaves. This monument leans northerly and projects more than usual; otherwise in good condition.

Monument No. 51, or Milestone XIX

This monument is situated in the woods, very near the top of the ridge between Negro and Shepherd ponds, on land belonging to the Pierson estate, and 60 paces from monument No. 52. At the monument the following bearing and distance were taken:

N 75° W, 3 feet to a small white oak, blazed. Soil sandy to within about six inches of the required depth, where a ledge of rock was encountered, on which a mark was cut in place of the disk. This monument (new one) has its southeast corner slightly chipped; otherwise in good condition. The old stone is an irregular trap-rock slab.

Monument No. 52, or Road Monument No. 1 between Milestones
XIX and XX

This monument is situated in the woods, about 60 paces west of milestone XIX and a little on the west slope of the ridge and on the east side of the road which runs N 20° E and S 20° W, and is about 180 paces southerly along the private road leading to F. L. Stetson's house from the main road, leaving the latter at the gate to William Hamilton's. At the monument the following bearings were taken: East 2 paces to a clump of three chestnut trees, and west 5 paces to another. Soil fine sand. Disk set in leaves. This monument has three of its corners chipped; otherwise in good condition. It should be moved to the new road near-by, as this road is abandoned.

Monument No. 53, or Road Monument No. 2 between Milestones
XIX and XX

This monument is situated near the top of the first hill west of Shepherd pond and on the east side of the old road running N 20° E and S 20° W which now terminates at a worm line fence about 1,800 feet from the main road. It is about 4,023 feet west of milestone XIX and on south side of worm fence. At the monu- ment the following bearing was taken: N 85° E to Hamilton's concrete house. Soil sandy. Disk set in wood ashes. This monument has its northeast corner slightly chipped; otherwise in good condition. It should be moved to the new road near-by, as this old road is abandoned.

Monument No. 54, or Milestone XX

This monument is situated in a wet meadow, about 170 feet west of the west edge of the woods which skirts the meadow, and is 5,227.4 feet west of milestone XIX and in a line between the lands now or formerly of Abram S. Hewitt and Colonel Payne, and southeast of a clump of bushes. At the monument the following bearing and distance were taken: S 58° E, 42 paces to a large white oak. Soil about two feet turf, two feet blue clay and the balance good solid sandy gravel; hole filled with water, but the monument is firmly set in cement and the disk put in as usual. This monument (new one) has its easterly corners slightly chipped; otherwise in good condition. The old stone is an irregular sandstone slab. This monument may be readily found by following the line fence westerly from monument No. 53.

Monument No. 55, or Road Monument No. 1 between Milestones XX and XXI

This monument is situated in the valley on the west side of a road which runs N 45° E and S 45° W. It is about 2,544 feet west of milestone XX and 5 paces north of a small culvert. At the monument the following bearing and distance were taken: N 65° E, 60 paces to the southeast corner of a frame house painted red. Soil gravelly and many boulders; hole filled with water. Disk set in wood ashes. This monument has its corners slightly chipped; otherwise in good condition.

Monument No. 56, or Milestone XXI

This monument is situated on the west slope of a high ridge. At the monument the following bearing and distance were taken: S 51° E, 4 paces to a white ash tree about 18 inches in diameter standing on line. Soil sandy. Disk set in leaves. This monument (new one) was found to be in good condition. The old one is an irregular granite slab.

To reach this monument take old woods road, Sterling Furnace to Ringwood, over the fence bars, past deserted house, over the bars again and around the hill.

Monument No. 57, or Road Monument No. 1 between Milestones XXI and XXII

This monument is situated on the west slope of the mountain, on the east side of a woods road from Ringwood to Sterling Furnace, which runs S 85° E and N 85° W, and about 80 paces west of milestone XXI. At the monument the following bearings and distances were taken: N 85° W, 7 paces to a white oak tree standing on the west side of the road, and N 60° E, 7 paces to a chestnut tree on the easterly side of the road. Soil sandy. Monument was broken in getting it to the place and in consequence was only set about two feet, but very firmly. This monument has three of its corners slightly chipped; otherwise in good condition.

Monument No. 58, or Milestone XXII

This monument is situated in the woods on the southeast slope of a hill which appears to be a projection from the Black Rock mountain. It is 5,258.46 feet west of milestone XXI and just on the southeast edge of a ledge of rocks. At the monument the following bearings and distances were taken: N 53° W, about 200 paces to the large boulder which caps the south end of Black Rock Mountain; S 48° E, about 4 paces to a small rock oak, and S 32° E, 16 paces to twin white oaks 10 inches in diameter, which separate about two feet above ground. Very rocky all about and in consequence a mark was cut in place of disk. This monument (new one) is chipped on the westerly edge; otherwise in good condition. The old one is an irregular granite slab the top of which is painted red.

The line has recently been cut and painted white; otherwise the monument would be difficult to find.

Monument No. 59, or Road Monument No. 1 between Milestones XXII and XXIII

This monument is situated in the valley between Black Rock and Beech mountains. It is about 4,012 feet west of milestone XXII and on the south side of a road which runs S 80° E. This monument is about 14 paces east of a small culvert. Solid rock was encountered and no disk used. This monument has its northeast and northwest corners slightly chipped; otherwise in good condition.

Monument No. 60, or Milestone XXIII

This monument is situated in a swamp, but on ground somewhat higher, about at the foot of Beech mountain, in thick woods, on land of F. K. Curtis. At the monument the following bearings and distances were taken: S 20° W, about 1,000 feet to the house formerly of William Patterson on Beech farm, and S 24° E, 25 paces to a large red oak. It is 4,586 feet west of milestone XXII and at the foot of two birch trees. Soil sandy. Disk set in leaves. This monument leans slightly to the northwest; otherwise in good condition. The old monument, a small, irregular slab, is not embedded and is one of the many stones surrounding the new monument.

Monument No. 61, or Milestone XXIV

This monument is situated on the west slope of Beech mountain, about half-way down, and on the west side of a road running about north and south, which at this point runs down hill towards the north. It is also 5,197 feet west of milestone XXIII. At the monument the following bearings and distances were taken: S 67° W, 40 paces to a large chestnut, which stands just on the east edge of a woods road that leaves the main road near the monument; about 3 paces northeast to three blazed trees on the west side of the road; also N 10° W, about 50 paces to a large white oak. Soil sandy. Disk set in leaves. This monument (new one) was found to be in good condition in every respect. The old stone is an irregular slab of mountain slate.

This line has just been cut through and painted white; hence it is easy to follow.

Monument No. 62, or Milestone XXV

This monument is situated on the west slope of the first ridge east of Greenwood lake and in the northwest corner of a swamp. It is also 5,247.4 feet west of milestone XXIV. At the monument, the following bearings and distances were taken: N 55° E, 5 paces to a large oak, and S 48° E, 60 paces to a large pepperage tree. Soil clay and somewhat wet. No disk used. This monument (new one) leans southerly; otherwise in good condition. The old stone is an irregular slab of sandstone. (Line opened and marked by white paint.)

Monument No. 63, or Railroad Monument No. 1 between Milestones XXV and XXVI, on The New York & Greenwood Lake Railway

This monument is situated at the west foot of the ridge on the west side of the New York & Greenwood Lake railroad track and is about 2,525 feet west of milestone XXV and 15 paces east of the edge of Greenwood lake. At the monument the following bearings and distances were taken: N 40° E, about 21 paces to the southeast corner of porch of hotel owned by Mr. Julius Brandes of Paterson, and S 40° W, 32 paces to the northeast corner of the railroad station. Soil sandy. Disk set in ashes from the railroad. This monument, 6 inches square, leans slightly to the southeast and has its corners and edges worn in consequence of its being in the center of a traveled roadway.

Monument No. 64, or Milestone XXVI

This monument, reset in concrete July 17, 1913, is situated at the east foot of Rough mountain, 5,282 feet west of milestone XXV and southeast 16 paces from the southeast corner of the Greenwood Lake Boat and Country Club house. At the monument the following bearings and distances were taken: N 23° W, 65 paces to a large sugar maple, which stands 10 paces west of the road, and S 70° W, 50 paces to another, which stands 10 paces east of the road and 5 paces northwest of boat-house of Greenwood Lake Boat and Country Club. Soil gravelly. Disk set in sand. This monument has three of its corners and one edge chipped; otherwise in good condition.

Monument No. 65, or Road Monument No. 1 between Milestones XXVI and XXVII

This monument is situated on the west side of the road which runs N 20° E and S 20° W along the east foot of Rough mountain and is about 50 paces west of milestone XXVI. At the monument the following bearings and distances were taken: N 12° E, 38 paces to a sugar maple, and S 5° W, 45 paces to another (both of which are mentioned in the preceding description), also 760 paces southwesterly along the road to the Lake Side house. Soil quite slaty. Disk set in charcoal. This monument has its corners and edges chipped; otherwise in good condition.

Monument No. 66, or Milestone XXVII

This monument is situated on Rough mountain about 300 paces west of the last or highest ridge and on a rocky ridge about midway between two swamps, and is distant 5,047.7 feet west of milestone XXVI. At the monument the following bearings and distances were taken: South 50 paces to a pine tree at westerly edge of east swamp; N 30° E, about 100 feet to the outlet of the easterly swamp; S 60° W, 7 paces to a small pine 8 inches in diameter growing out of the ledge, and N 30° W, 15 paces to small pine at northeast edge of west swamp. Cross cut on the ledge in place of disk. This monument was found to be in good condition in every respect.

To reach this monument take road from Bellvale running southwest between Bellvale mountain and Warwick mountain to gate marked " Cascade Park," which is about half a mile west of bridge over Long House creek. Pass through gate, follow woods road past abandoned house and orchard and over corduroy bridge spanning Long House creek, and continue past signs and along barbed wire fence and beyond to narrow crossing of swamp (25 feet, more or less, wide). Turn to right, follow westerly edge of swamp to outlet, cross outlet and follow description above. This crossing is within sight of the rocky ridge east of the swamp.

Monument No. 67, or Milestone XXVIII

This monument is situated about half-way up the southerly slope of a steep hill in the woods, 1 foot north of a barbed wire fence, 100 paces north of a cleared field and is distant 5,161 feet west of milestone XXVII. At the monument the following bearings and distances were taken: S 5° E, about 11 feet to a shell bark hickory; S 40° W, to an old tumble-down house standing on the north side of the lane which leads up by O'Brien's house, and S 45° E, 20 paces to a rock oak tree on line with barbed wire fence 200 feet, more or less, northwest of meadow up the hill. This monument (new one) has its westerly corner chipped; otherwise in good condition in every respect. The old stone is an irregular mountain slab, not embedded.

To reach this monument go in the lane south of road stone 68 feet and follow northeast to wire fence.

Monument No. 68, or Road Milestone No. 1 between Milestones
XXVIII and XXIX

This monument is situated on the west side of a road which runs N 20° E and S 20° W along the west slope of the ridge and is distant 1,295 feet west of milestone XXVIII. At the monument the following bearings and distances were taken: N 20° E, about 90 feet to a large chestnut standing in the stone fence; S 30° W, 7 paces to a large cherry tree, and 7 paces back along line to a twin oak on east side of road. Soil gravelly and some slate. Disk set in sand. This monument is chipped along its easterly and southerly edges and has three corners broken. It also leans slightly to the north.

Monument No. 69, or Milestone XXIX

This monument is situated about half-way down the westerly slope of a steep hill in an open field on land of Mr. Edward Wright, and is distant 5,232.8 feet west of milestone XXVIII. At the monument the following bearings and distances were taken: S 50° W, 50 paces to a large buttonwood tree, and N 19° W, 420 paces to the southeast corner of Wright's house. Soil sandy. Monument set in cement and disk in sand. This monument, new one, leans northwesterly; otherwise in good condition. The old stone is an irregular sandstone slab and lies near the new one.

Monument No. 70, or Road Monument No. 1 between Milestones
XXIX and XXX

This monument, reset July 15, 1913, is situated on the west side of a road which runs N 40° E and S 40° W and is distant about 1,813 feet west of milestone XXIX and is about at the west foot of the slope. At the monument the following bearings and distances were taken: N 50° E, 240 paces to a small house standing opposite the junction of the roads, also about 40 paces southwesterly along the road to a bridge, and 18 paces southwest to Paul Mezey's new house on west side of road. Soil sandy. Disk set in wood ashes. This monument has its northeast and southwest corners chipped; otherwise in good condition.

Monument No. 71, or Milestone XXX

This monument is situated on top of a mountain and on the east slope of a ridge, about 30 paces west of an old woods road, which runs between this ridge and one a little farther east, and 33 paces southeasterly along the line to the corner of a worm line fence. It is 5,325 feet west of milestone XXIX. At the monument the follow bearing and distances were taken: N 51° W, 7 paces to a large oak stump, and southerly 3 feet to a hickory sapling standing alone. Soil sandy. Disk set in leaves. This monument (new one) was found to be in good condition in every respect. The old stone is an irregular field-stone.

Monument No. 72, or Road Monument No. 1 between Milestones XXX and XXXI

This monument is situated on the east side of a road which runs N 40° E and S 40° W through the hollow, 25 feet east of middle of the road and 15 feet east of a brook. It is distant about 1,020 feet west of milestone XXX and 5 paces easterly of the point where the brook crosses the road and at the west foot of the hill. At the monument the following bearings and distances were taken: N 68° W, 20 paces to a large hemlock; S 65° E, 13 paces to a chestnut, also 3 paces to a twin black oak. Soil sandy, but very still and hard towards the bottom. No disk used. This monument has its southeasterly corner slightly chipped; otherwise in good condition.

Monument No. 73, or Road Monument No. 2 between Milestones XXX and XXXI

This monument is situated on the east side of a road running about north and south and about at the top of a hill and is distant 2,260 feet west of milestone XXX. At the monument the following bearings and distances were taken: N 8° W, 78 paces to a large hickory, now dead, which stands just on the roadside in front of the house of John W. House, and N 53° W, 22 paces to an apple tree standing in the orchard west of the road. Soil sandy. Disk set in wood ashes. This monument has its northeast and southwest corners slightly chipped; otherwise in good condition.

Monument No. 74, or Milestone XXXI

This monument is situated in an open field on the land of Frederick Carey, on the gentle west slope of the hill and is distant 5,280 feet west of milestone XXX and 22 paces west of stone wall on west side of lane leading past Carey's barn. At the monument the following bearing and distance were taken: N 44° E, 165 paces to the southeast corner of Carey's barn. Soil sandy. Disk set in sand and monument in cement. This monument, new one, has its northwest corner slightly chipped; otherwise in good condition. The old stone is gone.

Monument No. 75, or Road Monument No. 1 between Milestones XXXI and XXXII

This monument is situated on the west side of a road running about north and south and sloping towards the north and is distant 8,629.8 feet west of milestone XXX. At the monument the following bearing and distances were taken: N 7° E, 425 paces along the road to the house of Russell Ferguson, which stands at the junction of the road, and southerly 6 paces along the road to a red oak tree. Soil sandy. Disk set in wood ashes. This monument has its northerly and easterly edges slightly chipped; otherwise in good condition.

Monument No. 76, or Road Monument No. 2 between Milestones XXXI and XXXII

This monument is situated on the north side of a road running N 78° W and S 78° E, in front of and near the easterly end of a picket fence and between two shade trees, 6 feet 9 inches west of one and 19 feet 9 inches east of the other. At the monument the following bearing and distance were taken: N 35° W, 82 paces to the northeast corner of M. L. Taylor's house. Soil sandy. Disk set in wood ashes. This monument has three of its corners slightly chipped and leans a trifle to the northwest; otherwise in good condition.

Monument No. 77, or Milestone XXXII

This monument is situated in an open field belonging to M. L. Taylor, about 3 paces south of a lane running from his house down

across the Wawayanda, and is distant 5,317 feet west of milestone XXXI. At the monument the following bearings and distances were taken: S 34° E, about 75 paces to the northwest
corner of the main barn, and N 53° W, 110 paces to a large
hickory which stands in an open field along the aforesaid lane.
Soil sandy. Disk set in sand and monument in cement. This
monument, new one, has its northwest and southeast corners
slightly chipped; otherwise in good condition. The old monument
is of brownstone, well preserved and in good condition.

*Monument No. 78, or Railroad Monument No. 1 between Mile-
stones XXXII and XXXIII, on Lehigh & Hudson River
Railroad*

This monument, reset in concrete July 15, 1913, is situated
on the north side of the track of the Lehigh & Hudson River railroad and is distant about 2,367 feet west of milestone XXXII.
At the monument the following bearing and distances were taken:
S 88° E, 80 paces to an elm which stands in the field about 33
paces south of the track, also about 17 feet westerly along the
railroad to the northerly abutment of a cattle pass; soil sandy.
Disk set in ashes from the railroad. This monument, 6 inches
square, has three of its corners chipped; otherwise in good
condition.

*Monument No. 79, or Road Monument No. 1 between Milestones
XXXII and XXXIII*

This monument is situated on the south side of a road running
N 70° E and S 70° W, on ground comparatively level, and is
distant about 4,060 feet west of milestone XXXII. At the
monument the following bearing and distances were taken: N 35°
W, 25 paces to a butternut tree, also 263 paces southwesterly
along the road to the fork in the road. Soil slaty. Disk set in
wood ashes. This monument has all of its corners and easterly
and westerly edges chipped; otherwise in good condition.

*Monument No. 80, or Road Monument No. 2 between Milestones
XXXII and XXXIII*

, This monument is situated on the west side of a road which
runs N 25° E and S 25° W, at the east foot of a hill, and is

distant about 4,616 feet west of milestone XXXII. At the monument the following bearings and distances were taken: N 51° W, 4 feet to a large oak tree, now dead, on line; N 60° E, 45 paces to an elm, and east 25 paces to a black walnut tree. Soil sandy with slate rock bottom. Disk set in wood ashes. This monument has its two easterly corners and southerly edge chipped; otherwise in good condition.

Monument No. 81, or Milestone XXXIII

This monument is situated on the east slope of a ridge, just on the south side of a line stone fence on the property of A. Ely, now Mrs. Fairchild, at the top of a stone ledge, and is distant 5,355 feet west of milestone XXXII. At the monument the following bearings and distances were taken: N 51° W, 5 paces to a large white oak on line, and S 51° E, 10 paces to a large black walnut near line. Monument set in rock excavation with cross cut on the bottom in place of disk. This monument (both stones) was found to be in good condition in every respect.

Monument No. 82, or Road Monument No. 1 between Milestones XXXIII and XXXIV

This monument is situated on the west side of a road running N 50° E and S 50° W, and is distant about 4,354 feet west of milestone XXXIII. At the monument the following bearings and distances were taken: S 45° E, 12 paces to an apple tree near corner of a stone fence; S 50° W, 4 feet to a wild cherry tree; also about 3 feet southwesterly to line fence. Soil sandy. Disk set in wood ashes. This monument has its easterly corners chipped; otherwise in good condition.

Monument No. 83, or Milestone XXXIV

This monument, reset in concrete July 18, 1913, is situated in an open swampy field 200 feet west of west end of line fence on land of Mr. Layton and distant 5,280 feet west of milestone XXXIII and east of Pochuck meadow. At the monument the following bearings and distances were taken: S 5° W, 250 paces to the northwest corner of Layton's house, and N 51° W, 95 paces to a buttonwood standing on line. Disk set in fine gravel. This monument was found to be in good condition in every respect.

Monument No. 84, or Milestone XXXV

This monument, reset in concrete July 26, 1913, is situated in an open field on swampy low land about 100 paces east of the edge of the upland and the west edge of Pochuck meadow, 10 feet north of line fence, on the property of A. L. and F. L. Roy, and is distant about 5,347 feet west of milestone XXXIV. At the monument the following bearings and distance were taken: N 10° W, to Mr. Roy's house, and S 10° E, 6 paces to a pin oak tree standing on the south bank of a line brook, or ditch. This tree is not notched, or blazed. Soil soft and mushy. Disk set in sand and monument in cement. This monument (both stones) found to be in good condition in every respect except that new stone has its corners chipped.

Monument No. 85, or Road Monument No. 1 between Milestones XXXV and XXXVI

This monument is situated on the east side of a road running about at right angles to the line and on the east slope of the Pochuck mountain, about 9 feet north of line fence and 694.1 feet west of milestone XXXV. At the monument the following bearings and distances were taken: S 50° E, 43 paces to an elm tree standing in the fence, and S 24° W, 63 paces to another on the east edge of the road. Soil sandy. Disk set in sand. This monument has three of its corners chipped; otherwise in good condition.

Monument No. 86, or Milestone XXXVI

This monument is situated in an open field on the east slope of the Pochuck mountain, on the property of Daniel Bailey, and is distant 5,270 feet west of milestone XXXV. At the monument the following bearings and distances were taken: N 49° E, 40 paces to the southeast corner of the foundation of the old log house; S 64° E, 50 paces to a large willow tree; also about S 55° E, 36 paces to a black walnut tree, and N 50° E, 18 paces to an old railroad embankment. Soil sandy. Disk set in wood ashes. This monument projects a little more than usual; otherwise in good condition.

Monument No. 87, or Milestone XXXVII

This monument is situated in a ravine between two ridges of the Pochuck mountain, 30 feet north of the north edge of a swamp, on land of Jesse S. Lamareaux estate, and is distant 5,147.7 feet west of milestone XXXVI. At the monument the following bearings and distances were taken: S 51° E, 25 feet to an ash tree about 6 inches in diameter and N 75° W, 3 paces to a white oak standing about 30 feet south of the road leading to an orchard (following this road easterly about 100 paces it opens into an apple orchard). Soil sandy. Disk set in leaves. This monument (both stones) was found to be in good condition.

Monument No. 88, or Road Monument No. 1 between Milestones XXXVII and XXXVIII

This monument is situated on the westerly side of a road which runs N 43° E and S 43° W, about at the west foot of the Pochuck mountain, and is distant about 3,651 feet west of milestone XXXVII. At the monument the following bearing and distance were taken: S 46° W, 200 paces to the northeast corner of house now or formerly of Levine Potter. Soil sandy. Disk set in wood ashes. This monument has its easterly corner slightly chipped; otherwise in good condition.

Monument No. 89, or Milestone XXXVIII

This monument is situated in the meadow land east of the Wallkill river and distant 5,247.7 feet west of milestone XXXVII. At the monument the following bearing and distance were taken: N 53° W, 33 paces to a ditch running at right angles with the line fence. This monument is also in the line fence between the lands now or formerly of Mr. Lewis, and Drew and Home. Soil about one feet muck and the balance white sand. Disk set in sand, the monument in cement. This monument (new one) projects about 15 inches above the ground; otherwise it was found to be in good condition.

Monument No. 90, or Milestone XXXIX

This monument is situated on the east slope of and about 50 paces from the foot of the first ridge west of the Wallkill lowlands, in a stone fence on land of the Standard Oil Co. of N. J., and is

distant 5,278.7 feet west of milestone XXXVIII. It is also
about midway between the two westerly brick buildings of the
Standard Oil Co. of N. J., and at or near the southerly end of
same. At the monument the following bearings and distances
were taken: S 70° E, 48 paces to an oak standing just in the
edge of the lowland, and N 55° W, 5 paces to an oak stump standing just south of the fence. Disk set in wood ashes. This monument has three of its corners chipped and leans slightly to the
southwest; otherwise in good condition. We were unable to see
this monument, as it was covered with a coal pile ten feet high.

Monument No. 91, or Road Monument No. 1 between Milestones XXXIX and XL

This monument is situated on the west side of the road which
runs N 40° E and S 40° W, about at the west foot of a steep hill,
and is distant about 2,100 feet west of milestone XXXIX and
about 250 feet north of road running westerly. At the monument the following bearings and distances were taken: N 40° E,
425 paces along the road to house of the N. Y. Transit Co.; N 30°
W, to Mr. Clark's house, and N 60° E, 33 paces to a cherry tree.
Soil sandy. Disk set in wood ashes. This monument has three
of its corners chipped and projects more than usual; otherwise
in good condition.

Monument No. 92, or Road Monument No. 2 between Milestones XXXIX and XL

This monument is situated on the south side of a road which
runs S 50° E, about 100 paces along the road west of the top
and 13 paces west of a jog in the fence, and is distant about
4,173 feet west of milestone XXXIX. The road referred to
follows the boundary on the southwest side thereof nearly to milestone XL. At the monument the following bearing and distance
were taken: S 30° W about 250 paces to a house standing in a
hollow. Soil sandy. Disk set in wood ashes. This monument has
westerly corner slightly chipped; otherwise in good condition.

Monument No. 93, or Milestone XL

This monument is situated near the bottom of the west slope of a
hill in an old apple orchard, the land of Peter Kimber, and is dis-

tant 5,229.9 feet west of milestone XXXIX and about 30 feet south of road. At the monument the following bearing and distances were taken: N 78° W, 50 paces to a black walnut tree, also about 100 paces to the intersection of the road. Soil sandy. Disk set in sand and monument in cement. This monument (new one) was found to be in good condition in every respect. The old one, an irregular slab of slate, has split and has a portion broken off, lying beside it.

Monument No. 94, or Road Monument No. 1 between Milestones XL and XLI

This monument is situated on the west side of a road which runs nearly north and south, about 100 paces east of the west foot of a hill, and is distant about 278 feet west of milestone XL. At the monument the following bearings and distances were taken: about N 60° W, 100 paces to the northeast corner of Kimber's mill, and S 5° E, 70 feet to a black walnut tree. Soil sandy. Disk set in wood ashes. This monument has its easterly corners slightly chipped and leans slightly to the southwest; otherwise in good condition.

Monument No. 95, or Railroad Monument No. 1 (New York, Susquehanna & Western Railroad) between Milestones XL and XLI

This monument, reset in concrete July 29, 1913, is situated on the west side of the track of the New York, Susquehanna and Western railroad, about 3 feet from the westerly rail and just about at the north end of a slate cut, and is distant 500 paces west of milestone XL. At the monument the following bearing and distance were taken: N 45° E, 225 paces to the southeast corner of a barn. Disk set in wood ashes. This monument, six inches square, has three of its corners chipped, two of them badly; otherwise in good condition.

Monument No. 96, or Road Monument No. 2 between Milestones XL and XLI

This monument is situated on the east side of a road running N 25° E and S 25° W, is distant 2,122 feet west of milestone XL and is at west wall of mill-pond, also is about 20 paces

northeasterly from an intersection of the roads. Soil sandy for
one foot, but very hard and stiff. Slate in the bottom. Disk put
in the side and a mark cut on the rock. This monument leans
badly to the southeast, against the reservoir wall, has 3 feet exposed and has its northwest corner and westerly edge chipped. It
should be relocated across the road and set in concrete.

Extract from Mr. Hopper's report of 1916: " I was informed
that Mr. Wm. Vail, who built the wall of mill-pond on or about
1908, says he found the disk in the debris of material from wall
trench and hence there is a question as to whether or not this stone
has been shifted or changed from its original position."

Monument No. 97, or Road Monument No. 3 between Milestones XL and XLI

This monument is situated on the west side of a road which runs
N 35° E and S 35° W, about 75 paces from a fork in the road,
about at the west foot of a hill, and is distant about 5,083 feet west
of milestone XL. At the monument the following bearing and distance were taken: S 35° W, 65 paces to a chestnut tree standing
on the roadside at the top of the hill. Soil sandy. Disk set
in wood ashes. This monument has its easterly corners slightly
chipped; otherwise in good condition.

Monument No. 98, or Milestone XLI

This monument is situated in an open field on the west slope of a
small knoll, on land of Mr. Arriman, and is distant 5,280 feet west
of milestone XL. At the monument the following bearings and
distances were taken: N 45° E, about 200 paces to the southeast
corner of Arriman's barn, and S 5° E, 115 paces to the chestnut
tree mentioned in the preceding description. Soil sandy. Disk
set in sand and monument in cement. This monument projects
only 3 inches above the ground and has its westerly corners slightly
chipped; otherwise in good condition.

Monument No. 99, or Road Monument No. 1 between Milestones XLI and XLII

This monument is situated on the north side of a road which
runs N 68° W and S 68° E and is distant about 418 feet west of
milestone XLI. At the monument the following bearing and dis-

tance were taken: S 30° E, 240 paces to the chestnut tree mentioned in the description of monument No. 97. Soil sandy. Disk set in wood ashes. This monument has its corners chipped, two of them badly; otherwise in good condition.

Monument No. 100, or Road Monument No. 2 between Milestones XLI and XLII

This monument is situated on the east side of a road which runs N 25° E and S 25° W, on the westerly slope of a hill, and is distant about 2,906 feet west of milestone XLI. At the monument the following bearings and distances were taken: S 80° E, 12 paces to the northwest corner of R. E. Hallock's wagon-house, and N 35° E, 14 paces to a pear tree standing east of the road. Soil sandy. Disk set in wood ashes. This monument has three of its corners slightly chipped and leans a trifle northeasterly; otherwise in good condition.

Monument No. 101, or Road Monument No. 3 between Milestones XLI and XLII

This monument, reset in concrete July 30, 1913, is situated on the west side of a road which runs N 10° E and S 10° W and is at the east foot of a hill and just opposite or in line with the line fence between the lands of B. J. Hait and Charles Goldsmith. It is distant about 4,624 feet west of milestone XLI. At the monument the following bearings and distance were taken: S 10° W along the road to the northwest corner of Hait's house, and N 45° E, 145 paces along the road to Goldsmith's barn, standing on the east side thereof. Soil sandy. Disk set in wood ashes. This monument does not project as much as usual and has its northwesterly corner slightly chipped; otherwise in good condition.

Monument No. 102, or Milestone XLII

This monument is situated on the east slope of a hill, about 15 paces from the top, 3 feet north of a line fence between the lands of Hait and Goldsmith and in a stone wall running northerly therefrom. It is distant 5,261.95 feet west of milestone XLI. At the monument the following bearing and distances were taken: S 35° E about 300 paces to the northwest corner of Hait's house, and southeast 3 feet to large oak tree, now dead. Soil sandy, mixed with considerable slate. Disk set in wood ashes. This monument (both stones) was found to be in good condition in every respect. The old stone is an irregular slab of slate.

Monument No. 103, or Road Monument No. 1 between Milestones XLII and XLIII

This monument is situated at the west foot of a steep slope, on the west side of a road which runs N 50° E and S 50° W, and opposite a line fence, and is distant about 1,250 feet west of milestone XLII. At the monument the following bearings and distances were taken: S 51° E, 25 paces to a white oak tree; N 30° E, about 250 paces to the southeast corner of Beatty Brink's house, and S 51° E, 9 paces to a large crows-foot cut in the ledge of rock. Soil sandy. Disk set in wood ashes. This monument has three of its corners slightly chipped and leans slightly to the southwest; otherwise in good condition.

Monument No. 104, or Milestone XLIII

This monument is situated in a line fence between lands of Lewis Clark and John Bossler, in low ground, and is distant 5,270 feet west of milestone XLII. At the monument the following bearings and distances were taken: N 32° W, 57 paces to a sugar maple, and S 51° E about 66 paces to a wire fence running at right angles southwesterly from stone wall. Soil sandy. Disk set in wood ashes. This monument (both stones) was found to be in good condition in every respect. The old stone is an irregular slab. It was rather difficult to find this monument, standing as it does in the center of a stone fence on the line between lands of Lewis Clark and John Bossler.

Monument No. 105, or Road Monument No. 1 between Milestones XLIII and XLIV

This monument is situated on the west side of a road which runs N 30° E and S 30° W, is on the east slope of a small hill about 25 paces from the top, and is distant about 1,014 feet west of milestone XLIII. At the monument the following bearing and distance were taken: S 51° E, 9 paces to an oak on line, which stands just on the east edge of the road and is the corner between the townships of Greenville and Minnesink, also about 6 feet north of the corner of a line fence between lands of Lewis Clark and John Bossler. Soil sandy and very light. Disk set in wood ashes. This monument has all of its corners chipped, the northwest corner quite badly; otherwise in good condition.

Monument No. 106, or Road Monument No. 2 between Milestones XLIII and XLIV

This monument is situated on the west side of a road which runs about north and south, is on ground sloping gently westward and is distant about 3,685.9 feet west of milestone XLIII. At the monument the following bearings and distances were taken: S 51° E to the corner of a line fence between lands of Alice Northrop and Everett Forgerson; N 55° W, 300 paces to the northeast corner of Forgerson's house, which stands just south of a lane that leaves the main road 150 paces north of the monument, and N 10° E, 60 paces to a white oak tree standing by the roadside. Soil very rocky towards the bottom. Disk set in wood ashes. This monument was found to be in good condition in every respect.

Monument No. 107, or Milestone XLIV

This monument, reset in concrete July 31, 1913, is situated in a kind of low swampy meadow on the north edge of the lane which runs by Everett Forgerson's house under wire fence and about 20 feet east of the edge of the upland and is distant 5,304 feet west of milestone XLIII. At the monument the following bearings and distances were taken: S 51° E, 5 paces to an oak on line, and N 51° W about 30 feet to a chestnut, now dead, on line, and stone and wire fence running southwest from line. Soil clay with some gravel. Disk set in wood ashes. This monument (new one) was found to be in good condition. The old one is an irregular slab, which is firmly embedded, but inclines to the northeast.

Monument No. 108, or Road Monument No. 1 between Milestones XLIV and XLV

This monument is situated on the south side of a road which runs N 65° E and S 65° W, is on ground sloping gently eastward and is distant about 2,965 feet west of milestone XLIV. At the

monument the following bearing and distance were taken: S 20°
W, about 100 feet to a large maple tree standing on the east bank
of a brook and in a line fence between the lands of John Taylor
and Erastus Courtwright. Soil sandy. Disk set in wood ashes.
This monument has its northwest corner slightly chipped; otherwise in good condition.

Monument No. 109, or Milestone XLV

This monument is situated in the northeast corner of a meadow
belonging to F. Braisted, just at the southwest corner of the woods
and nearly in line with worm fence running southerly, and is distant 5,301 feet west of milestone XLIV. At the monument the
following bearings and distances were taken: N 55° W, 450 paces
to the northeast corner of Braisted's house, which stands on the
west side of the road running along the east foot of Blue mountain,
and N 44° E, 15 paces to the line fence referred to in the description of monument No. 110. Soil clayey. Disk set in wood ashes.
This monument (new one) projects about two feet above the
ground, otherwise it was found to be in good condition. The old
one is an irregular slab.

Monument No. 110, or Road Monument No. 1 between Milestones XLV and XLVI

This monument is situated on the east side of a road which runs
N 25° E and S 25° W along the east foot of the Blue mountains
and is distant 47 feet south of the corner of a line fence between
the lands of John Gilson and F. Braisted. At the monument the
following bearings and distances were taken: S 55° W, 34 paces
to the northeast corner of F. Braisted's house, and N 25° E, 200
paces to the southeast corner of Gilson's house. Soil sandy. Disk
set in wood ashes. This monument was found to be in good
condition in every respect.

Monument No. 111, or Milestone XLVI

This monument is situated in scrub oaks on property of Ayer,
on the west slope of the east summit of the Blue mountains, about
200 paces from the summit. At the monument the following
bearings and distances were taken: N 50° W, about 18 feet to a

small blazed pine tree, and S 50° E, 50 paces to two large boulders about on line. Soil sandy. The monument having been broken in getting it to the place was set in only about 1½ feet, but was very firmly wedged. Disk set in wood ashes. This monument (both stones) was found to be in good condition in every respect. The old stone is an irregular slab.

This stone is difficult to reach without a guide. The route runs north of, up and around a steep hill back of Braisted house till you come to Ayer's wire fence and a woods road. Follow road over corduroy crossing of brook, then to the left along another brook and through swampy ground toward High point to "Keep off" sign, where marked trees on line begin. Follow these up over hill to edge of swamp and continue to monument, as described.

Monument No. 112, or Milestone XLVII

This monument is situated on the west slope of the Blue mountains in low thick brush on land of James Hamilton, and is distant 5,280 feet west of milestone XLVI and about 5 paces northwest of the edge of a woods road bearing S 65° W. At the monument the following bearings and distances were taken: S 40° W, 10 paces to a pine oak tree which stands near the angle in the road, also N 40° W, about one-third of a mile to James Hamilton's house. Soil sandy. Disk set in wood ashes. This monument (new one) was found to be in good condition in every respect. The old monument is gone.

Monument No. 113, or Road Monument No. 1 between Milestones XLVII and XLVIII

This monument is situated on the east side of a road which runs a little northeast and southwest and at the foot of Hogback mountain and is distant about 10,000 feet west of milestone XLVI. At the monument the following bearing was taken: N 47° W to the south end of Alex. Burrow's house, or saloon, which stands about 4 feet clear in New York. The line at this point passes between his house and saloon, also on or near the line fence between lands of Michael Fitzsimmons and W. H. Vail. Soil sandy. Disk set in wood ashes. This monument is covered and could not be found. It should be raised and set in concrete.

Monument No. 114, or Milestone XLVIII

This monument is situated on the top of the Hogback mountain in a line wire fence, now down, between lands of Thomas Dutton and the widow Snyder. Also it is on the edge of a path, is 40 feet east of a ridge of rock and 40 paces west of an oak tree about 10 inches in diameter, which stands in said line fence, and is distant 10,419 feet west of milestone XLVI. At the monument the following bearing and distance were taken: N 30° E, 15 paces to a pine tree. No disk used. This monument (new one) has its northwest and southeast corners slightly chipped and leans slightly to the southwest; otherwise in good condition. The old one is an irregular slab and leans slightly to the northeast.

Monument No. 115, or Road Monument No. 1 between Milestone XLVIII and Neversink River Monument

This monument is situated on the west side of a road which runs N 25° E and S 25° W, is at the easterly foot of a slope, and is distant about 300 paces west of milestone XLVIII. Also it is on the line fence between Snyder and Rutan and 26 inches outside the fence in the road. At the monument the following bearing and distance were taken: N 30° W, 100 paces to the southeast corner of Rutan's house. Soil sandy. Disk set in wood ashes. This monument has three of its corners slightly chipped and is about one inch below the surface of ground at the side of the road; otherwise in good condition. As other monuments in the vicinity should be raised and set in concrete, this one may well be reset at the same time.

Monument No. 116, or Neversink River Monument

This monument is situated about 15 paces east of the easterly water edge of the Neversink river and in the line fence between lands of C. W. Rutan and the estate of A. P. Snyder, and about 1,310 feet from monument No. 115. At the monument the following bearing and distance were taken: S 50° E, about 10 feet to a locust tree. Soil sandy. Disk set in wood ashes. This monument (six inches square) has its easterly corners and northerly and southerly edges chipped; otherwise in good condition.

Western Witness, or Reference, Monument

This monument stands on an eminence about midway between the Delaware and Neversink rivers and on land of the Laurel Grove Cemetery Co., and is N 64° E, 72¼ feet from the Tri States rock, or monument. It is similar in form and dimensions (above ground) to the witness, or reference, monument at the eastern terminus of the line and, in addition to the inscription cut on that monument, it is further marked on one of its edges with the words "Witness Monument" and on the north side with the words "The corner between New York and Pennsylvania is in the center of the Delaware river, 475 feet due west of the Tri State Rock" and on the New Jersey side with the words "South 64 degrees W. 72¼ feet from this is the Tri State Rock which is the northwest end of The New York and New Jersey boundary and the north end of the New Jersey and Pennsylvania boundary." This monument has the northeast corner of its base and the southwest corner of the shaft and base slightly chipped, otherwise in good condition, except that the southwest side of base needed to be reinforced with fresh masonry, which was supplied on September 8, 1916.

Tri States Monument

The point which this monument is intended to define was originally indicated by a crow's-foot cut in the natural limestone rock and which in 1874 was very plain, although its cut edges were somewhat smoothed by the exposure of 100 years. In 1874 the United States Coast and Geodetic Survey, at the request of the Geological Survey of New Jersey, determined accurately the latitude and longitude of this point, and at the close of the work marked it by drilling a deep hole in the rock and fastening in it a copper tube filled with lead and setting and describing proper witnesses of its location. The station point according to this determination is in latitude 41° 21′ 22.63″ north, longitude 74° 41′ 40.70″ west from Greenwich.

By order of the joint commission on boundary line between the states of New York and New Jersey the copper bolt was excavated October 30, 1882, in making foundation for the existing granite monument, the center of which was placed directly over the point occupied by the bolt. The monument as originally set in 1882 was similar in form and dimensions and built into the solid rock in the same manner as the witness, or reference, monument at the eastern terminus of the line.   In the spring of 1883 the upper portion was broken off by ice and on May 21, 1885, by order of Commissioner Leavenworth of New York and Commissioner Cook of New Jersey the remaining portion was redressed to the existing dimensions, which are as follows: 2 feet 4 inches long, 1 foot 4 inches wide, and 1 foot 5 inches high above the surface of the rock in which it is embedded.   Upon its top surface it was marked with ¼-inch grooves, showing the directions of the lines of the three states which meet there, and within the surfaces bounded by the line, the initials of the respective states are cut.   The north side of the stone is further marked with the words " Tri States Monument."   This monument is built into the natural rock at the junction of and near the extreme high-water mark of the Delaware and Neversink rivers.   It was in good condition, as above described, except that its southeasterly perpendicular edges were slightly chipped and the southwest side needed to be reinforced with fresh masonry, which was applied on September 8, 1916.   The joint inspection of 1919 found it in first-class condition.

Notes

While the above descriptions refer directly to the new monuments, erected in 1882, it should be understood that they also refer indirectly, as regards location, to the old original monuments, which in 1882 were first reset in a substantial manner in their original location, the new mile monuments being then set on the east side contiguous to them and in line therewith.

The disks referred to in the above descriptions are of earthenware, 6 inches in diameter, 1 inch thick and perforated in the center and are, unless otherwise stated, placed vertically beneath the point to be marked by the monument and 6 inches beneath its bottom.

As showing the relative value of the different monuments we quote from an act entitled,

"An act to ratify and confirm the agreement entered into by Commissioners on the part of the States of New York and New Jersey in relation to that portion of the boundary line between said states extending from the Hudson river on the east to the Delaware river on the west.

" SECOND. The monumental marks by which said boundary line shall hereafter be known and recognized are hereby declared to be: First, the original monuments of stone erected in 1774, along said line by the Commissioners aforesaid as the same have been restored and re-established in their original positions by Edward A. Bowser, surveyor, on the part of New Jersey, and Henry W. Clarke, surveyor, on the part of New York, duly appointed by the parties hereto; Second, the new monuments of granite erected by the aforesaid surveyors at intervals of one mile, more or less, along said line and numbered consecutively, beginning from the Hudson river, and severally marked on the northerly side with the letters ' N. Y.' and on the southerly side with the letters ' N. J.'; and Third, the monuments of granite erected by the aforesaid surveyors at intervening points on said line at its intersection with public roads, railroads and rivers, and severally marked by them, on the northerly side with the letters ' N. Y.' and on the southerly side with the letters ' N. J.'; and Fourth, the terminal monuments erected at the western terminus of said line at the confluence of the Delaware and Neversink rivers, and the terminal monuments erected on the brow of the rock called the Palisades near the eastern terminus, and the rock lying and being at the foot of the Palisades on the bank of the Hudson river, and marked as the original terminal monument of said line established in 1774; as the same are described in a joint report made to the parties hereto by Elias W. Leavenworth, Commissioner on the part of New York. and George H. Cook, Commissioner on the part of New Jersey."

INDEX OF ILLUSTRATIONS

INDEX

Annual Report of the State Engineer and Surveyor of New York, 1919

E

M

N

O

Lightning Source UK Ltd.
Milton Keynes UK
UKHW010001090219
336872UK00005B/288/P